Quantitative Risk Management and Decision Making in Construction

Other Titles of Interest

Water Engineering with the Spreadsheet: A Workbook for Water Resources Calculations Using Excel, Ashok Pandit, Ph.D., P.E. Provides more than 90 problems, from easy to difficult, in the areas of fluid mechanics, hydraulics, hydrology, and stormwater management, ISBN 978-0-7844-1404-0

Slope Stability Analysis by the Limit Equilibrium Method: Fundamentals and Methods, Yang H. Huang, Sc.D., P.E. Presents fundamental principles and methods for using the limit equilibrium method in analyzing slope stability for the safe design of earth slopes, 2014, ISBN 978-0-7844-1288-6

Entropy Theory in Hydraulic Engineering: An Introduction, Vijay P. Singh, Ph.D., P.E. Explains the basic concepts of entropy theory from a hydraulic perspective and demonstrates the theory's application in solving practical engineering problems, 2014, ISBN 978-0-7844-1272-5

Risk and Reliability Analysis: A Handbook for Civil and Environmental Engineers, Vijay P. Singh, Ph.D., P.E.; Sharad K. Jain, Ph.D.; and Aditya Tyagi, Ph.D., P.E. Presents the key concepts of risk and reliability that apply to a wide array of problems in civil and environmental engineering, 2007, ISBN 978-0-7844-0891-9

Modeling Complex Engineering Structures, Robert E. Melchers, Ph.D. and Richard Hough, eds. Provides an overview of cutting-edge developments in computational theory and techniques as currently applied in various fields of structural analysis, in the United States and around the world, 2007, ISBN 978-0-7844-0850-6

Quantitative Risk Management and Decision Making in Construction

Amarjit Singh, Ph.D., P.Eng., C.Eng.

ASCE PRESS

Library of Congress Cataloging-in-Publication Data

Names: Singh, Amarjit, 1952- author.
Title: Quantitative risk management and decision making in construction /
  Amarjit Singh Ph.D., P.Eng., C.Eng., PMP, FICE, FASCE, FCIOB.
Description: Reston, Virginia : American Society of Civil Engineers, [2017] | Includes bibliographical
  references and index.
Identifiers: LCCN 2016044518 | ISBN 9780784414637 (soft cover : alk. paper) | ISBN 9780784480298 (PDF)
Subjects: LCSH: Construction industry–Management–Data processing. | Risk management–Mathematical
  models. | Decision making–Mathematical models.
Classification: LCC TH438 .S527 2017 | DDC 624.068/1–dc23 LC record available at https://lccn.loc.gov/
  2016044518

Published by American Society of Civil Engineers
1801 Alexander Bell Drive
Reston, Virginia 20191-4382
www.asce.org/bookstore | ascelibrary.org

Dedicated to

The One and Only One on High,
Who Alone is Without Fear, Beyond Uncertainty,
and Above Risk,
For Whom Nothing is Unknown

Contents

Preface

Risk is a matter of perennial interest. Every large living organism has a nervous system capable of sensing risk in order to make all decisions about advancement and survivability, as evident by the instinct for survival in times of danger. However, one's senses do not always live up to the demand, perhaps mostly because some risks are more daunting than the mind is capable of processing and handling. Moreover, risks often arrive without warning, as if at random, which makes risk analysis genuinely challenging. Hence, an improved risk assessment system is needed to help decision makers in production, construction, business, and organizational management.

Risks exist simply because the forces of nature and life do not fall on a deterministic scale that humans can discern. The science of events and why they happen is still largely unknown. For instance, floods and hurricanes strike unpredictably. When the economy will improve or worsen, if and when an armed rebellion will take place, or how and in which fashion laws affecting the daily lives of citizens will change are not issues for which ordinary humans have any crystal ball.

In fact, risk pervades our daily lives: Crossing the street, driving a car, investing in stocks, catching a flight, and buying unguaranteed products are laden with risk. Taking any type of medicine carries risks of side effects, and eating ordinary food carries the risk that it may not be good for the body. Simply breathing can carry health risks if one lives in a heavily polluted area, and even taking a stroll in the park or walking on the grounds of a university carries the risk of stepping on uneven ground and twisting one's ankle.

Political risk is even more challenging to estimate and can be disturbing for many who struggle with its consequences. There is more of the unknown in evaluating political risk than in industrial, manufacturing, or construction risk. Often, the methods of activation of political risk are beyond scientific determination, border considerably on the realms of chaos, and are a function of psychology and individuality that are difficult to estimate in their own right.

However, engineering risk has more defined borders. For instance, the limiting factors are usually materials, labor productivity, equipment, design specifications, and personnel quality. All of these can be mitigated by intelligent management and control mechanisms and effective data collection and analysis. Even some of the more adverse risk factors, such as weather disturbances and labor strikes, are not altogether without control measures. Furthermore, insurance plays a great role in assuring that losses as a result of engineering risks are mitigated, whereas there is little practice of such insurance for political risk, although insuring shipping during military engagements or threats of military intervention has been undertaken, such as during the various Persian Gulf crises.

Among the most important considerations in engineering and construction risk is the effect it has on project price, because the bottom line of any engineering or business decision is the likely impact on revenue, expenses, and final profit. This is a serious matter for any small business and cannot be ignored by large agencies and corporations. Although there are ways to ensure that the bottom line is not affected—by focusing on multiple management, customer, quality, and safety parameters—the hard reality of business and engineering is predicated on financial success, for which risk analysis is basic and fundamental.

For all risk analysis, it is important to understand the basic paradigm of *known knowns, known unknowns,* and *unknown unknowns,* explained in Chapter 1. This sets the stage for our work on risk analysis and helps us focus on issues with direct impact on decision making. One can manage and attempt to control only the known knowns and known unknowns, so struggling with unknown unknowns is usually not a profitable way of managing the project or business—although a small contingency fund can, no doubt, be set aside for handling risks that were not thought of at the time of project estimation.

This book is written with the purpose of introducing valuable techniques of weighing and evaluating alternatives and items that cause risk in decision making. Common strategies and techniques are covered. Bear in mind that the effect of a risk is activated only by the type of decision made to counter that risk. Not making well-informed decisions can end in losses and regrets. Hence, addressing decision making and how to make optimal decisions is the purpose of this book.

The focus of this text is on quantitative risk-analysis techniques that have been developed in industry over the past several decades. The common techniques presented can be executed for business and construction risk assessment needs without requiring high-level mathematics, which business and construction managers are not likely to use anyway. Hence, the focus is on bid risk, evaluating contingency funds, identifying causes and effects, and diagnosing them by structuring their "how" and "why." We will proceed to examine how to assign weights, test their accuracy, and then use them for effective utility decisions; we will then move on to more precise decision-making techniques of pay-off matrices, optimizing for inventory control, and considering control limits for effective production management. Decision analysis, which always requires input of mental energy for evaluating existing conditions and alternatives, is presented via decision-tree analysis, which builds on cause and effect and the FAST diagram. A special chapter on educational risk management is presented that evaluates how to win a card game about risk. In this card game, all the 2.64 million possible card combinations have been analyzed to arrive at a conclusion about an effective strategy for deciding whether to "build" or "not build."

The further purpose of this book is to offer a comprehensive text for graduate and senior undergraduate students. All thirteen chapters in this book are what I cover in the one-semester, three-credit course on quantitative risk analysis that I have taught since 2006 and have covered in other classes before that. I pulled together into one course all the quantitative decision making and risk analysis topics I taught. Hence, this text covers exactly what can be feasibly covered over a single semester, nothing more. Depending on how fast the course instructor wishes to go and the skill and talent level of the students, the instructor using this book may wish to teach only eleven or twelve chapters, not all thirteen. The instructor would be encouraged to add variations to tailor the course to the taste of students and instructor alike.

A course using this text can be taught at the graduate level, senior level, or mixed senior and graduate level. Such a course is found to particularly benefit students with a management bent of mind and those who like to solve problems of uncertainty.

Each chapter comes from a larger body of knowledge and has been selected for inclusion owing to its use in practical decision making. Chapter 1, "Risk Management Planning," is taken from the material on risk management that is covered by the Construction Industry Institute. Chapters 2, "Probabilistic Cost Analysis," and 3, "Contingency Analysis and Allocation," are part of the larger study area of cost engineering and cost estimating. The material in Chapters 4, "Cause-and-Effect Diagrams," and 6, "Decision Trees," is heavily used in quality management and utility theory. Chapter 5, "Function Analysis System Technique Diagrams—Structuring Uncertainty," is extracted from value engineering, which is an independent field in its own right and has enormous potential for generating improvements in virtually anything to which the human mind applies itself. Chapter 7, "Payoff Matrix," comes from decision-making theory, which helps greatly when managers need to make a decision in accordance with their specific management style and aversion to risk. Chapter 8, "Bayes' Theorem," is usually taught in statistics and has applications in forensic engineering. Chapter 9, "Matrix Analysis," is an extension of value engineering and is excellent for ranking alternatives. Chapter 10, "MCDM and the Analytic Hierarchy Process," has mathematical origins but has been immensely popular in many fields for evaluating the weights managers usually assign to priorities and criteria before making a decision. Chapter 11, "Project Planning—The OOPS Game," can be considered as part of gambling theory, but it should be noted that the OOPS System and the programs and strategies developed in this book will be found nowhere else. Chapter 12 is basic to quality control, in which production managers seek to catch risk as soon as they can so as to avoid losses later. Chapter 13 takes a page from inventory management and inventory control with the aim of reducing disturbances in inventory management, given that materials management is fundamental to construction. All of these chapters have their homes in diverse fields of study and engineering applications.

The author received help with the various chapters from engineers who are recognized and acknowledged here. Jeremy Adaniya assisted the author with the chapter on risk planning. The chapter on probabilistic estimating has been rewritten based on one of my articles that was published in the journal *Cost Engineering;* Alvin

Magallanes helped in that rewrite. All of the chapters rely on my personal notes and collections over the years, which I share with my graduate students in the course. Bao Lin made contributions to the chapter on contingency analysis, and Kari Kumashiro helped with cause-effect analysis. Alvina Lutu worked hard on FAST diagramming and displayed patience through all the revisions. Stephen Peters basically wrote the chapter on decision trees, and Benjamin Card wrote about payoff matrices—both under my strategic guidance and direction. Alvin Magallanes put together material on Bayesian analysis. David Bushnell supplied material about matrix analysis and satisfaction factors. Rhandi Ardona helped with the analytic hierarchy process, discovering new things in the process. James Wessel wrote the computer program and helped crack the "OOPS!" game; I owe sincere thanks to Greg Howell for introducing me to it. This is a great game for educating students about risk, in addition to getting them interested in learning about risk and motivated on the subject. James Wessel wrote the chapter, subject to my reviews and revisions. After Megan Inouye and I worked on a journal paper using control charts, she very adeptly put together the chapter on tracking performance. Garon Nobriga contributed to inventory risk management, patiently working with me to make revisions. Maja Caroee prepared the first appendix on creating an actual risk plan, and Bao Lin conceived the problem presented in Appendix 2. Finally, I owe a tip of the hat to those who helped me format the manuscript: Sophie Friedheim, Sofia Oliveira, and Karen Toyoda. Manuela Melo helped me repair figures and find vital references during the copyediting and proofreading stages, in addition to helping me check the proofs. Alvin Magallanes further helped in checking and verifying the calculations in the analytic hierarchy process and economic order quantity segments.

I thank all those mentioned for their patience in acquiring material for me and putting together chapters and sections of chapters according to my demanding requirements, specific needs, and challenging scope, often under time constraints. Essentially, I am indebted that they put up with me; without them, I could not have reached the conclusion of this work.

In addition, my interaction with Martin Loosemore was what got me moving on this book, after I spent a sabbatical with him at the University of New South Wales in Sydney, teaching quantitative risk analysis. Seeing that he had only then penned a book on risk management, I saw that there was scope in the industry for a book on the quantitative methods used for assessing risk and making decisions. Martin is a great writer himself: If not for the ideas and inspiration I got from talking to him and spending quality time together, this book would not have seen the light of day.

CHAPTER 1

Risk Management Planning

1.1 Introduction

Uncertainty, along with death and taxes, will occur in life. Whether it involves day-to-day activities or significant large-scale activities, there will always be uncertainty, or risk, that must be taken into account. Risk is the measurement of the potential for a loss. In addition, this loss is directly proportional to the variance of the variable for which risk is being measured; the larger the variance is, the larger the risk. For example, the weather in the midwestern United States during tornado season can be much riskier than the weather in southern California during the spring.

However, risk is not completely negative, because opportunity—the measurement of the potential for a gain—opposes risk. The often subjective relationship between a risk and the associated opportunity depends on the situation, context, individual, organization, and culture. For example, two individuals may perceive the same risk as providing different opportunities because of their perception or experiences. Most risks and their associated opportunities generally have a constantly changing relationship, depending on the circumstances and people involved. Consequently, when studying risk and its opportunities, one must minimize subjectivity, promote collaboration, and vacate assumptions to gain a thorough and analytical set of information.

The planning of a construction project must consider risk. However, all too often project teams, which may include government officials, engineers, and other professionals, ignore or pay insufficient attention to risks that are inevitable in construction projects. The consideration of uncertainty in a project, or *risk management planning*, benefits all parties involved in a project and can significantly enhance the project's success by ensuring timely and cost-effective completion. As a process, risk management planning requires the recognition, understanding, and resolution of all potential risks throughout the course of a project (Construction Industry Institute 1989). The risk management plan for project organization documents and plans this process and must include the following six steps:

1. Risk identification,
2. Risk assessment and analysis,
3. Risk control,
4. Risk cost allocation,
5. Risk monitoring, and
6. Risk planning.

Large or complex projects—as well as new or previously unattempted projects—have a greater need for such plans. Developing a thorough risk management plan requires detailed and accurate documentation throughout the process. Consequently, a properly developed risk management plan allows a project team to confidently and effectively execute a project with minimal delays and cost overruns. This chapter aims to give the reader a complete understanding of the risk management planning process and its execution.

1.2 Risk Identification

The management of risk first requires a project team to collaboratively identify and categorize a project's risks, such as those related to safety, performance, and business. To identify the potential risks, team members can analyze project documents and use their collective experience and various techniques, including brainstorming (possibly with sessions on a daily basis). Categorizing the risks can involve multiple rounds of categorization, and evaluation meetings can help the team organize the risks and create risk checklists and databases for current or future projects (Ashley et al. 2006).

There are three types of risks: *known, known unknown,* and *unknown unknown. Known* risks are those already understood to exist and certain to occur or affect the project. This type of risk occurs frequently and can include minor variations throughout the project. An example of a known risk is price swings in materials costs. A *known unknown* risk is one with an unknown probability of occurrence and unknown effect on the project. An example of this type of risk is extreme weather—hurricanes, floods, etc. An *unknown unknown* risk is an unforeseeable risk not yet materialized from which unimaginable risks can arise. Although they are unrelated to construction, AIDS and Ebola epidemics exemplify unknown unknowns which, before their discovery, were unforeseeable and not experienced by anyone. Once an unknown unknown materializes, it ceases to be an unknown unknown and definitely becomes one of the known unknowns; with additional exposure, it can move into the realm of the known. Knowledge of these different types of risks can aid a project team in conducting a thorough and thoughtful risk identification process.

1.2.1 Risk Recognition

As the first step of the risk identification process, the project team can recognize a project's potential risks by thoroughly examining elements related to the project's scope. This method tends to be the most appropriate, because these elements typically describe the entire body of work that the project team needs to address (Le et al. 2009). The project documents contain the scope of the project and its related elements, such as

- General contract conditions,
- Technical specifications,
- Cost estimates,
- Special conditions of the contract,
- Project description,
- Historical data about the project or area,
- Design and evaluation reports,
- Technical studies, and
- Risk checklists.

Using these documents, the team can use its collective experience and employ techniques such as brainstorming, scenario formulation, and expert interviews to begin the process of recognizing and creating a list of the project's risks.

For example, some of the risks associated with a parking-lot construction project could be determined based on a project document such as an archaeological report. The United States has special construction laws related to the discovery of archaeological sites, which means the possibility should be taken into account at the planning stage. The project team could review this report and brainstorm possible related risks, such as the potential risk of delays owing to, say, time spent extracting and examining artifacts properly. That same archaeological report could also suggest a risk that the project's costs will increase because of, say, artifact extraction costs. Thus, the same event can have multiple risks, and the project team must recognize the associated risks and must not mistake the event itself for the risk. This situation exemplifies the process that the project team should undertake during the risk identification process. The team would then create a list of associated risks and must ensure that the list of the project's risks has enough detail to ensure that the team's members and stockholders clearly understand the risks now and later.

1.2.2 Risk Categorization

After identifying the potential risks, the project team will categorize them. Organizing and categorizing the risks eases its management later in the process. Furthermore, categorizing risks helps reduce redundancy and, as mentioned previously, provides the foundation for creating risk checklists and databases for future projects.

The first step of the categorization process is to filter out risks deemed unimportant by the project team. To do this, the team must first determine whom the identified risks affect, directly or indirectly, whether it be the contractor, the owner, both, or some other party. Understanding who is affected can narrow the project team's focus to only relevant risks.

The next step in the categorization process is to group risks according to their sources, which could include

- Performance issues,
- Quality issues,
- Environmental concerns,
- Safety issues,
- Cost uncertainty,
- Schedule uncertainty,
- Staffing and manpower, and
- Political concerns.

Risks can be grouped by the sources from which they emanate, and the project team can develop checklists that categorize various risks according to their respective sources. Going back to the parking-lot construction project example, the risk of delays resulting from the discovery of ancient artifacts would fall within the category of scheduling risks. Similarly, the risk of increased project costs because of the need to remove these artifacts would belong to the category of cost risks.

Once the risks' sources have been identified and the risks placed into their appropriate category, each category can be further categorized as *internal* or *external*. An internal risk is a controllable risk for the project team, such as scheduling or costs. An external risk is an uncontrollable risk for the project team and could include community groups, regulators, or others. External risks tend to be unpredictable, yet foreseeable, such as the risks associated with the uncovering of ancient artifacts; these risks can be essentially understood as known unknowns. Once this categorization process has been applied, the project team will have formulated a diverse and clear list of the risks and their categories. Although additional categorization can occur, this level of categorization is generally sufficient, and the team can move on to analyzing the identified risks.

Although it is considered the first step of risk management, risk identification is also a continuous, iterative, and evolving process throughout the entire project. The identification process sets the foundation for the risk management process; thus, potential risks must be recognized and categorized as early as possible to reduce or eliminate future issues. Accordingly, successful risk identification requires a group effort; the more experience and perspectives are shared, the greater the likelihood that all potential risks will be identified.

1.3 Risk Assessment and Analysis

The next step in the Risk Management Plan is risk analysis—the process of quantifying the identified risks. The analysis of risk requires the determination of each risk's likelihood and severity. Determining these two characteristics is difficult, and some risks (e.g., an environmental or public health issue) may be hard to represent with a discrete number or value. Nevertheless, the use of subjectivity when analyzing risk can be problematic and should be avoided when possible: A well-reasoned estimated value, even if difficult to determine or assign, is far more beneficial than no value at all.

Various procedures and techniques can be used for risk analysis. A complete and thorough risk analysis process includes both qualitative and quantitative analysis of all risks. In addition, a thorough risk analysis should assess all possible sources of risk (Skorupka 2008). Qualitative analysis is useful for analyzing risks and categorizing them by their impact. Furthermore, as mentioned, sometimes a risk analysis cannot be represented well quantitatively, because of either the type of risk or a lack of information. In such cases, a qualitative analysis can be used instead, in order to represent a risk's value more thoroughly. A quantitative analysis is useful in making comparisons and providing an actual value for a risk, which can be used in the risk control step. Nevertheless, *it is best practice to complete both qualitative and quantitative analyses of all risks* to help in assigning appropriate risk control procedures and to create a complete risk management program to mitigate risk.

1.3.1 Qualitative Risk Analysis

Qualitative risk analysis usually serves as the most useful step in the process by laying the foundation for further analysis, whether quantitative or some other type. The qualitative analysis helps team members to understand the risks in real terms, which is essential for the execution of all project plans. A qualitative analysis requires the identification, assessment, and assortment of the probability and effects of potential risks (Smith et al. 2006). Sometimes this type of analysis is the only type that can be performed because of a lack of information; however, qualitative analysis is frequently sufficient to provide enough information to determine risk control strategies. Nevertheless, a thorough analysis will include as much information as possible and should include both qualitative and quantitative approaches.

The first step in qualitative risk analysis is the determination and classification of the severity and likelihood of a risk. Various sources can contribute to this determination, such as

- Professional and personal experience,
- Historical information,
- Academic research,
- Model projections, and

- Analysis and evaluation of various types, including statistical, technical, managerial, political, or other ratings.

The classification of a risk's severity can also be represented in various ways (Ashley et al. 2006):

- No impact,
- Minor impact,
- Moderate impact,
- Major impact, or
- Critical impact.

These impacts could affect costs, quality, or other elements of the project or be measured in total; they can be assigned numerical values if required.

Similarly, the degree of likelihood that a risk will happen can be classified as

- Unlikely,
- Remote,
- Reasonably possible and likely,
- Very likely, or
- Certain to near certain.

Once the severity and likelihood of a risk occurring are determined, the project team can then categorize the various risks as low, moderate, or high. This process is performed within the categories determined previously in the risk identification step so that the low, moderate, and high risks within each category are clearly represented—as mentioned, for risk planning is an iterative process. A low risk has little to no impact and a low likelihood of occurring. A moderate risk has a high severity but a low likelihood of occurring, or a low severity but a high likelihood of occurring. Finally, a high risk has both a high severity and a high likelihood of occurring. Determining these categories for each risk helps the project team decide which risks deserve the most attention. Fig. 1-1 includes two charts that describe the different levels of a risk's severity and likelihood, which then lead to a grid that displays the different possible combinations. These different combinations determine whether each risk is low, moderate, or high. Note that, by this stage, the risk is already well structured.

A risk's category—low, moderate, or high—determines the protocol used to handle that risk. Often a low risk is disregarded, as it does not warrant further attention. The project team should review it periodically and may elevate it to a higher risk category if necessary. There are various techniques for handling moderate risks. If the moderate risk is of low severity but high likelihood, it tends to have a minimal impact on the project. Often the project team considers the effects of these types of risks combined with other moderate risks of the same type

Level	Likelihood
A	Unlikely
B	Remote
C	Reasonably possible and likely
D	Very likely
E	Certain to near certain

Level	Severity
P	No impact
Q	Minor impact
R	Moderate impact
S	Major impact
T	Critical impact

Severity

Likelihood	P	Q	R	S	T
A	L	L	L	L	M
B	L	L	L	M	M
C	L	L	M	M	H
D	L	M	M	H	H
E	M	M	H	H	H

Risk Category
H: High risk
M: Moderate risk
L: Low risk

Fig. 1-1. Qualitative risk analysis diagram

before determining how to manage them (Ashley et al. 2006). Such risks may include material prices, scheduling changes, and other considerations. Conversely, the project team manages risks of high severity with low likelihood on an individual basis; such risks may include technical complexity or material shortage, among others. The project team actively manages risks of this type throughout the project to track their probability and severity. Last, high risks are also closely managed on an individual basis. The project team should give risks of this type its utmost attention and make them its highest priority in order to mitigate losses and project failures. For risks categorized as moderate or high, the project team applies risk control measures after the risk analysis step. Thus, the project team should develop a risk prioritization system.

Qualitative risk analysis is the first and often the most important step in risk analysis and management (Smith et al. 2006). It requires the project team to classify the identified risks, within the categories developed during the identification process, by their severity and likelihood of occurrence and then further categorize them to determine which risks require the most attention. Only then can the team go through each category and identify the moderate to high risks, so that it can apply the appropriate risk control measures.

1.3.2 Quantitative Risk Analysis

A thorough risk analysis will evaluate, both qualitatively and quantitatively, a project's potential risks. The qualitative analysis provides a nonnumerical assessment of the probability and effects of the project's risks; this assessment then aids in categorizing the risks to determine which risks require the most attention. Sometimes this analysis provides sufficient information for the team to decide on the necessary risk control measures. In addition, quantitative information for a risk is not always available, in which case a qualitative risk analysis alone will have to suffice. Nevertheless, when possible, a quantitative analysis should be conducted on the important risks identified through qualitative analysis. Quantitative analysis provides a value for a risk's overall impact on the project, based on its severity and likelihood. This value further supplements the available data, which the project team can use to more thoroughly determine what risk control measures to apply and how.

Various methods can be applied in the quantitative analysis step. However, it helps to first understand the three types of quantitative risk analysis in construction management: technical performance analysis, schedule risk analysis, and cost risk analysis (Ashley et al. 2006).[1] The technical performance analysis addresses the question of whether a project will work by analyzing equipment, materials, and other factors. The schedule risk analysis, as the name suggests, analyzes the project's schedule and whether it will be completed on time. Finally, the cost risk analysis examines the risk associated with the project's various costs, such as labor, materials, and cost overruns.

Understanding the types of risk analysis used can aid the project team in narrowing the choice of available methods. Certain methods tend to be more effective for certain types of risk analysis; thus, the project team should consider this when selecting which method to use. Methods that can be used for quantitative analysis may include, among others,

- Value assignment,
- Simulation,
- Probabilistic techniques, and
- Influence diagrams.

1.3.2.1 Value Assignment

The project team performs a value assignment, usually a subjective method in which the team uses its collective experience and historical data to determine a numerical value for a risk's impact on a project. Typically, this method is used for analyzing technical performance and cost risks. Once the project team identifies the important

---

[1]It is noteworthy that schedule and cost are primary risk concerns for the construction owner. Nevertheless, multiple factors for ensuring success are important to keep in perspective (refer to Ashley et al. 1987, Baker and Murphy 1988).

risks and completes the qualitative analysis, it can perform a value assignment to evaluate each risk's impact on the project. This value allows the team to compare a project's risks numerically to include each factor. Furthermore, this numerical value can aid in assigning contingency amounts for each risk in the risk control step.

Value assignment benefits from its simplicity and ability to help assign contingencies to risks; however, it suffers because of the subjectivity required. Moreover, value assignment ignores the analysis of a risk's cause, which can lead to an incomplete evaluation of that risk. For example, for a risk of a change in the cost of materials, value assignment assigns the same value to the risk whether the cause is major, such as a change in supplier, or minor, such as small, random price fluctuations. Similarly, with this method the risk's value does not reflect an analysis of the risk's consequences; a value assignment applies the same value to a risk despite varying potential consequences. Using the same materials cost example as given, a value assignment would apply the same value to the risk whether the consequence was a loss of $1,000 or $1 million. This method of quantitative analysis may suffice in a number of cases, but the project team must avoid oversimplifying the risk and creating too many generalities, which can lead to poor risk management and damaging consequences.

1.3.2.2 Simulation

Simulation is common in practice and most effective when applied to risks associated with the project's costs and schedule. Simulation methods use computer probability models and probability distributions to predict a range of outcomes for the project owing to a particular risk—for example, the Monte Carlo method uses objective numerical values derived from data for its analysis. This helps avoid the problems associated with subjectivity in risk analysis and provides for extensive and impartial data concerning a risk's impact on the project. Although this method is very effective and informative, it requires extensive training to implement correctly. Furthermore, it requires specific information on the probability distribution of a risk, such as its mean, standard deviation, and general shape. This information, unfortunately, is not always available for a particular risk; thus, simulation cannot always be implemented successfully.

1.3.2.3 Probabilistic Techniques

Probabilistic techniques use various mathematical probability calculations to determine a value and range for a risk's impact on a project. They require only a few numerical values for the risk, such as the mean and standard deviation of that risk's probability, yet they still allow for low subjectivity in the determination. These methods accommodate some subjectivity and adjustment of the calculations by the project team if necessary; however, it can still provide a reasonably objective value for a risk's impacts, as well as the contingency amount, if needed, for the risk control step.

One of the disadvantages of probabilistic techniques, though, is that they do not provide much information beyond the value of a risk's impact. Compared to simulation, probabilistic techniques are less effective at providing the information necessary to aid the project team in making well-formulated decisions for a risk control strategy. Also, probabilities can be difficult to apply to certain risks, especially those associated with a project's schedule. Probabilistic techniques, however, can still effectively provide an objective analysis of a risk's impact using very limited information, as long as the project team understands the shortcomings.

1.3.2.4 Discrete Events

The discrete event method typically uses probability trees, decision trees, or influence diagrams. These diagrams display the sequence of multiple connected events, decisions, or influences and can be applied to all types of risks. However, they require extensive expertise and input data. The diagrams can present quantitative data pertaining to a risk, such as probability or cost. As the diagrams branch out, they display the consequential value of that risk's impact. At each consecutive step in the diagram, branches represent the different potential values of the previous step's impact. Thus, at their conclusion, the diagrams display all possible outcomes and values of a particular risk's impact. At that point, each path to a particular outcome on the diagram is given a score to which all other outcomes on the diagram can be compared and which can aid the project team in determining what risk control measures to apply (Lifson and Shaifer 1982). Because this method requires extensive training and data, it is commonly used for the most difficult projects. It effectively represents a risk and its impact's value.

This section has presented four methods for conducting a quantitative risk analysis. Choosing the most appropriate method requires the project team to understand each method's advantages and disadvantages, as well as to determine the type of quantitative risk analysis being conducted—whether technical performance, schedule, or cost–risk analysis. In addition, the method that a project team chooses depends on the available input data, the required outputs and detail, and the project team's available experience and skill sets. Only when these factors are considered thoroughly will the project team be able to complete its quantitative risk analysis successfully and effectively apply appropriate risk control measures in the following steps of the risk management process.

1.4 Risk Control

Following the risk identification step and the subsequent risk analysis, the next step in the Risk Management Plan is risk control, which represents the natural progression of a project's response to a potential problem. Risk control explores risk response strategies that can address those risks identified as of high importance, which include

- Risk reduction,
- Risk distribution, and
- Risk acceptance.

The strategies follow from the presumption that at the end of the risk analysis, the project team must have a grasp on risk. Each strategy has its strengths and weaknesses, and some strategies can be more effective than others, depending on the situation. For each risk identified as of high importance, the project team must consider all risk control strategies before choosing the optimal one. The project team should also consider a comprehensive risk control strategy, which considers all risks and possible strategies, in order to minimize multiple types of losses resulting from risks.

1.4.1 Risk Reduction

Risk reduction is a risk control strategy that aims to eliminate, or at least reduce, a project's risks. It is often the easiest strategy for the project team to implement first, because it is completely within the team's control and requires little to no additional funding. The project team must determine how extensively it wishes to reduce a risk while maintaining the project's goals and benefits, because reducing a risk can often affect these outcomes.

The project team can implement the strategy of risk reduction to completely avoid an identified risk. This strategy may require a project to be discontinued, especially when analysis indicates that losses resulting from the risk strongly outweigh the benefits and profits. Another instance of risk elimination can occur when the project team changes the project's plans and uses a method with lower risks, thereby eliminating the previous risk by changing various aspects of the project, such as its scope, schedule, technology, and available resources.

Using the earlier example concerning the construction of a parking lot, the risk of delays caused by the uncovering of ancient artifacts can be eliminated by relocating the construction site to another place that has no chance of possessing artifacts. Although eliminating a risk can simplify the risk management process, eliminating a risk by changing a project's plan can also potentially decrease the benefits of a project. Typically, the greater the risks in a project plan, the greater the possible benefits. Thus, by using an alternative and less risky project plan, the project team may reduce the project's benefits. The significant choice of eliminating a risk, therefore, requires careful decision making.

Rather than eliminating a risk, a less drastic form of risk reduction is simply mitigating the risk by degrees. The project team can accomplish this by finding alternatives, such as a change in method, equipment, supplier, or schedule. If there is a risk of delay—e.g., because of those ancient artifacts, for example—the risk can be reduced by creating a schedule that allows other construction to continue uninterrupted while the artifacts are managed. This strategy would prevent the complete stoppage of the project and could reduce delay. The alternatives used in this form of risk reduction would not be as extensive as those used in eliminating a risk. Consequently, the possible loss of benefits owing to reducing the risk would be less significant than the loss resulting from completely eliminating it.

Nevertheless, implementing the risk reduction strategy can be time-consuming and costly; it may require additional effort and analysis by the project team to develop alternatives that reduce or eliminate the risk while minimizing the benefits lost. The project team must also analyze how an alternative can impact the overall project. Thus, before applying the risk reduction strategy, the project team must determine whether the potential benefits of risk reduction outweigh the time and cost involved.

1.4.2 Risk Distribution

Risk distribution is a risk control strategy that distributes the risk among the parties involved with the project, thus reducing or eliminating the risk for a single party. One valid reason for risk distribution is that the risk is outside the area of expertise of the main party holding the risk. In some cases, the parties may share the risk; in other cases, the risk can be completely transferred from one party to another. Because this risk control strategy involves multiple parties, it can be more complex and difficult for the project team to implement. Therefore, the project team should examine all risk reduction strategies before implementing risk distribution.

One type of risk distribution strategy, the partial distribution of an identified risk, can serve as a strategy when the owner and the contractor create a joint venture or use a project plan that shares the risk between the two parties. For example, the contractor responsible for constructing the parking lot in the previous example could create a contract with the owner specifying that if ancient artifacts are discovered, both parties must agree to adjust the project's schedule without any adverse effects to either. Sharing a risk motivates all parties to monitor and address it throughout the project. However, involving additional parties is the primary challenge of risk sharing. It can be difficult to persuade one party to manage a particular risk, especially because other parties within the project are also trying to reduce their own risks and may resist taking on more. In these situations, negotiation and bargaining may be used to ensure that all parties involved have their interests addressed.

Another form of risk distribution is completely transferring a risk from one party to another. For example, the project team may wish to apply this strategy when hiring subcontractors. This strategy can eliminate a risk identified as of high importance by the project team and simplify the risk management process for the team. As with sharing a risk, finding a party willing to accept the entire risk could be difficult, but there will often be a price at which the party will agree to assume a risk. In the example of subcontractors, using multiple subcontractors may allow the project team to completely distribute multiple risks to other parties. Using a subcontractor for only a small aspect of the project offers the subcontractor with a smaller task with fewer risks, possibly making that party more willing to accept the contract. However, the project team must ensure that all parties are aware of and can handle the risks with which they are dealing. Otherwise, the complete distribution of a risk may be more hazardous than not distributing the risk at all (Flanagan and Norman 1993).

Furthermore, even if a risk is distributed completely to another party, the project team must still consider other risks attached to that risk. For example, a project team may decide to completely distribute the safety risks associated with demolition by contracting the task to a demolition company. However, the team will still have to consider other risks associated with those safety risks, such as demolition scheduling risks and environmental risks. Furthermore, the project team must closely analyze the cost of completely distributing a risk, compare it to the cost of taking on that risk either completely or partially, and determine which situation benefits the project the most.

1.4.3 Risk Acceptance

Risk acceptance is a risk control strategy in which the project team accepts a risk. Typically, this acceptance comes with a backup plan, such as insurance or a contingency. However, there are instances when the project team may accept a risk without such a plan or contingency. Risk acceptance often requires funding on behalf of the involved party in order to implement the plan properly, and thus it is often the last risk control strategy to be considered, after assessing the risk reduction and risk distribution strategies. When deciding whether to accept a risk, the project team must carefully consider and compare risk acceptance strategies to one another and to the other risk control strategies to make the optimal decision.

Risk acceptance with insurance is a strategy used when the project team accepts a risk and decides to protect itself from the risk's associated losses by purchasing insurance. Typically, this strategy aims to protect the project from the unknown unknown type of risk. The project team could use self-insurance, a form of contingency; however, this can be hazardous and expensive. The conventional form of insurance is obtained from an external source, such as a commercial or state agency, with a straightforward cost—workers' compensation, property damage liability insurance, or others. When using an external insurance agency, the project team needs to consider deductibles for its insurance plan. The deductible tends to represent the amount of self-insurance that the project team is willing to accept when a risk is realized but carries risks of its own. Thus, it should be factored into the risk management process and undergo analysis as well, for it too may require risk control measures.

Risk acceptance with contingency is a strategy in which the project team accepts a risk and creates its own fund—a contingency—to cover the potential losses. The contingency is a reserve account that can hold either money or time. Various methods can help the team determine the multiple aspects of the contingency, such as its amount, allocation, and usage. The Monte Carlo method, previously mentioned for its use in risk analysis, can be used to calculate a contingency and its many aspects. Risk acceptance with contingency is typically used for known and known unknown risks or for small, repetitive risks. The project team should consider other risk control strategies before accepting a risk and using its funds to create a contingency, because this strategy can carry its own risks. If the contingency is too high, it could lead to poor cost management; if it is too low, it may create a rigid and unrealistic financial environment for the project (Creedy et al. 2010). However, certain situations or risks may require acceptance with contingency because of other costs or constraints. There are instances, nonetheless, when a project team may benefit from accepting a risk and its contingency, in case the risky event does not happen. A contingency is complex and has its own risks and benefits, which the project team must analyze carefully, along with the project's overall risks and benefits, to maximize the benefits.

Risk acceptance without a contingency involves the acceptance of a risk without any backup funds to cover the associated losses. A project team should consider risk acceptance only after considering all other risk control strategies, and it should only be applied to those risks with a low impact and low potential for occurrence. Although these low-impact risks typically do not require the use of a risk control strategy, there may be instances when the project team considers such a risk significant enough to require one. In addition, the project team should consider this type of risk without a contingency only when the handling of those risks influences the overall decision about a project. For example, a project may seem viable and beneficial after the project team has applied risk control strategies to all but a few risks, which, though small and infrequent, may yet be important. In this case, the project team could consider accepting those few risks without a contingency if the benefits of the overall project outweigh the potential losses from those small risks. Accepting a risk without a contingency has its own risk and should be done with caution.

The project team must be aware that the risk control step of risk management is a complex and iterative process, with risk control measures carrying their own risks. A less-than-optimal risk control strategy may be applied to a particular risk if it contributes to the project's overall benefits and goals. If a thorough study and application of risk control strategies is shown to maximize the project's benefits, then the risk management process can progress to the next step of allocating risks.

1.5 Risk Allocation

After risk control, risk allocation aims to allocate the costs of risks to the appropriate parties, often by assigning risks to the parties deemed most capable of managing them and applying the risk control strategy. Risk allocation is an important process in risk management, presenting the early stages of structuring the risks and their control measures into an overall plan for risk management. However, there may be projects for which such an allocation strategy minimizes the risk but fails to maximize a project's benefits. In this case, the project team must determine the project's priorities:

minimize the risks or maximize the benefits. Based on these priorities, the team can then allocate and share the risks appropriately. For example, a public-works project may be needed sooner than would be possible under typical allocation guidelines. In this case, the project team will recognize the project's benefit—to serve the public—as the most important priority; the risks are allocated or shared to fulfill this goal first, even if doing so increases risks or costs. The project team must thoroughly consider the project's priorities, because those priorities can significantly affect the allocation of risks and the project's outcome if a typical allocation strategy is not effective.

1.6 Risk Monitoring

The development of a monitoring program aims to track the risks identified in the project's risk management plan and identify any new risks, whether caused by changes in the project or inaccuracies in the original plan. This is usually done after allocation of the risks originally identified. The monitoring program lasts throughout the project and often leads to changes to the risk management plan.

The monitoring program should regularly check and compare the risk management plan with the project to ensure that the project and its risks are progressing as planned. It also includes periodically reviewing those risks. Sometimes this periodic review requires the project team to go through the entire risk management process to ascertain that the project's risks are relevant and current. Usually review is necessary when a new risk develops because of a change in the project or inaccuracies in the original risk management plan. Furthermore, the project team must evaluate whether established risks are affected or if these new risks affect the overall plan. This may require repeating the risk management process for all risks and adjusting the overall plan.

In addition, the project team should develop a comprehensive reporting procedure for these monitoring activities. The reporting system should thoroughly document the progression of the project and its risks, as well as any changes to the overall risk management plan and the project team's decisions. The risk management plan must be periodically monitored; any negligence in adhering to the plan can result in failure to achieve project aims and objectives.

Either a whole department or the project manager alone can monitor the risks, depending on the size of the project, the complexities faced, and the uncertainties encountered. The monitoring program—a continuous process—is the only aspect of the risk management process that accounts for the constant flux a project encounters.

1.7 Risk Planning

The final step in the risk management process—risk planning—uses information gathered from the previous steps to develop an organized and comprehensive risk management strategy. Conducting the previous steps, a project team will have identified the project's risks and developed control strategies as well as developed an allocation and monitoring program for those risks.

The planning process assembles that information into an overall structure that defines the project's risk management plan. This structure is developed primarily through the scheduling of occurrences and actions related to the project's risks, as described in the previous steps, including the scheduling of a risk's occurrence, its control strategy, its allocation, and its monitoring. The project team can schedule this information to minimize the project's risks and maximize the benefits. This scheduling prepares the project team for any situation without affecting the project's progress, thus saving time and costs. Scheduling a project's risks and control strategies should result in a timeline that guides how a project and its risks would progress under ideal circumstances. Most often, however, the project does not progress as planned, but the planning process sets a reference from which adjustments to the project and its risks can be designed.

In addition, the project team can address the allocation of the project's risks in the planning process. Although a risk can be allocated to a particular party throughout the project, there may be times when the allocation of the risk needs to change. For example, if the nature of the project or risk, as planned, changes during the course of the project, then the party best able to absorb the risk may change as well. Thus, planning a project's risk allocation aids in addressing and preventing this potential issue. The planning process can also be applied to the project team's monitoring program, as the team defines the method for monitoring and reporting the project's risks. The planning process can supplement this information with a schedule stating when the monitoring plan should be implemented for specific risks. Certain risks, because of their nature, must be monitored more frequently. The planning process can also determine when the project's risks should be evaluated to see if any new risks have developed. This may entail frequent evaluation periods during stages of the project with more risks, or the schedule may stay consistent throughout the project.

The planning process plays a significant role in defining an overall structure for the risk management plan. It uses the information gathered during the risk management process to assemble a comprehensive and thorough risk management plan that aims to minimize a project's risks while fulfilling its goals. Once the planning step has been completed, the risk management process can conclude.

1.8 Summary

At the end of a risk management process, the project team can finally create a risk management plan—a document with information gathered in six steps: risk identification, risk assessment and analysis, risk control, risk allocation, risk monitoring, and risk planning. First, risk identification identifies and categorizes the potential risks associated with a project. Second, risk analysis analyzes the risks qualitatively and quantitatively to identify those risks of high importance. Third, risk control applies various strategies to maximize the project's benefits and minimize losses resulting from those risks. Fourth, the allocation process allocates the risks to the appropriate parties. Fifth, a monitoring program is developed to track the progress of the project and its risks and update the risk management plan whenever a new risk develops. Finally, the risk management plan concludes with the planning of possible risks, risk control strategies, allocation, and monitoring. Only after all these steps have been completed and information gathered and documented can the project team create the overall risk management plan, which provides a comprehensive plan for managing a project's risks. After that, the entire project team responsible for managing the risks should, preferably, agree and sign off on the risk management plan.

A continuous, evolving, and often iterative process, risk management requires vigilance and thoroughness by the project team. Although the process can be tedious and time-consuming, risk management can save significant costs and time while offering substantial success in project completion. In many cases, unfortunately, agencies, companies, and groups ignore these important facets of risk management planning, or they may give up on the creation of a risk management plan altogether because of the amount of work involved. However, because many projects need to be completed in less time and with less money, risk management can ensure their timely completion and increase productivity and overall satisfaction.

1.9 Exercises

Problem 1

You are the project manager of a charity organization that is planning to construct a small two-lane bridge spanning a stream in Nicaragua. The monsoon period is May to November, so it is unlikely that any construction can take place then. The bridge is small, only 30 ft in length, and its height above the stream bed is 10 ft. The water depth during the construction season (December to April) varies from 2 ft to 6 in. It is expected that construction should not take more than six weeks. The project consists of concrete abutments to be constructed on either side, as well as precast beams that must be manufactured off-site and trucked to the stream. But the site is along an unpaved dirt road in a hilly region 10 miles from the nearest city. Local workers will be used, which will be arranged by the local contractor.

You must make sure that the villagers, local cooperative, mayor, and other local groups remain committed to and in favor of the project. You need to manage the local contractor, although you are from the United States. You don't speak Spanish, but you can engage a few students who do to help you in translating and in coordinating the project. All told, you are on a tight budget, so the contractor needs to give you a guaranteed price while delivering a project that meets good construction practices and standards. Soil testing, professional design, and shop drawings have all been undertaken and prepared. You are tasked with the project delivery.

You are in charge of creating the project's risk management plan. Meet with your team and create a basic risk management plan by completing the following tasks. Feel free to establish other conditions and constraints as your group thinks fit.

1. Identify and categorize a few potential risks for the project.
2. Apply qualitative risk analysis procedures to the identified risks.
3. Determine proper risk control strategies for the risks.

Problem 2

Using the information given and your solution to Problem 1, conclude the risk management planning process by completing the following tasks:

1. Allocate risks appropriately to the parties involved with the project.
2. Devise a monitoring program for your risk management plan.
3. Prepare a crisis management plan.
4. Devise an overall plan for your project's risk management.

References

Ashley, D. B., Lurie, C. S., and Jaselskis, E. J. (1987). "Determinants of construction project success." *Project Manage. J.*, **18**(2), 69–79.

Ashley, D. B., Diekmann, J., and Molenaar, K. (2006). "Guide to risk assessment and allocation for highway construction management." Rep. FHWA-PL-06-032, Federal Highway Administration, Washington, DC.

Baker, B. N., and Murphy, D. C. (1988). "Factors affecting project success." *Project management handbook*, 2nd Ed., D. I. Cleland and W. R. King, eds., Van Nostrand Reinhold, New York, 902–919.

Construction Industry Institute. (1989). "Management of project risks and uncertainties." RS 6-8, Austin, TX.

Creedy, G. D., Skitmore, M., and Wong, J. (2010). "Evaluation of risk factors leading to cost overrun in delivery of highway construction projects." *J. Constr. Eng. Manage.*, **136**(5), 528–537.

Flanagan, R., and Norman, G. (1993). *Risk management and construction*, Blackwell, Oxford, U.K.

Le, T., Caldas, C., Gibson, G., Jr., and Thole, M. (2009). "Assessing scope and managing risk in the highway project development process." *J. Constr. Eng. Manage.*, **135**(9), 900–910.

Lifson, M. W., and Shaifer, E. F. (1982). *Decision and risk analysis for construction management*, Wiley, New York.

Skorupka, D. (2008). "Identification and initial risk assessment of construction projects in Poland." *J. Manage. Eng.*, **24**(3), 120–127.

Smith, N. J., Merna, T., and Jobling, P. (2006). *Managing risk in construction projects*, 2nd Ed., Blackwell, Oxford, U.K.

CHAPTER 2

Probabilistic Cost Analysis

2.1 Introduction

This chapter will introduce a method for bidders to determine a bid price with confidence and without cost overrun. Because of the intrinsic uncertainties and cost variations of work packages that make up a project, the probability of underrunning and overrunning a bid price are finite. An example problem that is demonstrated later in this chapter uses the technique of evaluating bid contingency using triangular distributions, shows how the owner can use the project's cumulative probability profile to estimate bid price, and discusses risk transfer and risk absorption strategies.

Triangular price distributions are most preferred for their ease and practicality. Contractors are used to getting three prices on projects, so it is natural to use triangular distributions.

To decide on the precise amount to bid may be quite a demanding task for a contractor, in large part because of the risks that come with projects and bid items. These risks fluctuate for every project expense, can range from significant to insignificant, and may result in cost overruns or profit opportunities, as well as allowing the contractor to come out even.

There is no doubt that project expense items can be optimistic enough that a contractor can afford either to bid low or to place the standard bid and receive high returns. Conversely, costs can be pessimistic enough that the contractor is obligated to bid high and face lower profits, along with the risk of not winning the contract.

In order to submit a reliable estimate—and before contractors decide on a potential venture—contractors need to first analyze the risks and profitability of the project. If after analysis, the risks and profitability do not fulfill the contractor's expectations, it is unlikely that the venture will be pursued. But at what bid price should a contractor quote, what is a reasonable bid, and at what price is the venture justified?

This chapter focuses on answering these questions by explaining how to analyze these cost variations and apply the results to determine an appropriate bid price. In addition, it explains how to use the subsequent opportunities for cost underrun or overrun that may exist to determine a reasonable bid price. This would also enable an owner to determine reasonable contingency funds for potential cost overruns that a contractor should keep for a rainy day.

Use of risk analysis techniques have been abundantly beneficial, as demonstrated on various projects, largely because major construction projects can suffer from substantial cost overruns and delays owing to nonconsideration of risk. Formal risk assessments and analyses are now a requirement for construction projects in the public sector: The U.S. Federal Transit Administration requires a formal risk assessment for all new-start projects, and the Washington Department of Transportation has required a formal risk analysis for all capital improvement projects since 2002 (Reilly et al. 2004, pp. 53–75). A survey of CEOs of 25 megaconstruction firms showed that they believed that "properly managing and pricing risk is their biggest challenge" (Rubin 2005; Touran 2006), thus revealing the importance of risk analysis in any sector of construction. A report to the Construction Management Association of America stated that "risk analysis is a valuable source of information" for the construction manager and owner (Touran 2006). Thus, having a good grip on risk and uncertainty is a necessity in construction.

Molenaar (2005) described case studies of the actual application of programmatic risk analysis to nine highway megaprojects in the United States, with a mean total value of more than US$22 billion, indicating that risk analysis is taken seriously in some parts of the construction industry. The fact that probabilistic risk analysis was undertaken at all on so large a scale is remarkable.

2.2 Triangular (Three-Point) Estimates

Projects typically obtain three estimates for every cost item—from materials, equipment, and labor to any subcontractor price. It should also go without saying that estimators need to understand the methods involved in construction in order to determine an accurate estimate. Knowing the scope and specifications allows their fulfillment. Eventually, having three estimates helps ensure that costs can be limited and that the prices are representative and reliable, enabling the contractor to make an informed final decision.

As the contractor receives more estimates for subsequent purchases or projects, these estimates should be organized and subdivided into three ranges at which the contractor would likely be bidding: (1) a most likely range; (2) a pessimistic range; and (3) an optimistic range. Because of the amount of uncertainty and risk associated with costs, subdividing into ranges is additional work that does not yield much beneficial information. Hence, considering a range of three prices is both simple and representative.

Notice that PERT[1] calculations are also made with three-point estimates. Therefore, we see that triangular distributions have a history of application in science and can be

---

[1] Probability Evaluation Review Technique (PERT), a popular risk assessment duration and project scheduling tool.

put to use in the development of quantitative risk analysis for construction projects, as described in this chapter.

2.3 Cost Items (Work Packages)

Large construction projects, such as coal and nuclear power plants, usually comprise hundreds or thousands of work packages. As a simple demonstration of the technique of risk and contingency analysis, let us consider a project that is composed of three work packages, A, B, and C. Examples of work packages are foundations, civil works, and structural construction. Costs—such as overhead, personnel, and indirect costs—can also form part of these work packages, depending on how the costs are packaged.

2.4 Work Package Cost Fluctuations

Variability in project activities, materials, and labor costs, among other issues, can influence the probability of a cost overrun or underrun on a project. These issues can include, but are not limited to, absenteeism, labor sickness, injuries, worker tiredness, low worker morale, equipment breakdown, material shortages, delays in material delivery, faulty materials, schedule compression, work redoes resulting from poor workmanship, weather, natural disasters, crisis situations, drawing errors, owner interference, differing site conditions, and poor management and cost control, as well as other possible eventualities. A more elaborate list of risks on construction projects is given by the Construction Industry Institute (CII 1989).

2.5 Price Probability

Suppose that for Work Package A, the contractor receives only three price quotes: $5, $7, and $10 (in thousands of dollars). The probability of achieving a $5,000, $7,000, or $10,000 price on this project with a discrete distribution of one count each is 0.33 for each price, as shown in Fig. 2-1. If the contractor receives two quotes of $5,000, three quotes of $7,000, and one quote of $10,000, the probability of achieving each price is 0.33 for $5,000, 0.50 for $7,000, and 0.17 for $10,000, as shown in Fig. 2-2. Such a procedure for evaluating probabilities is valid when histograms are used to represent the frequencies of prices.

Grouping the data into suitable class sizes can be convenient, because the prices of different vendors, suppliers, and subcontractors are usually unique. Thus, four ranges of $4,000 to $5,099, $6,000 to $7,099, $8,000 to $9,099, and $10,000 to $11,099 might be suitable in this case. Other groups can be created for the estimator's convenience. Such histograms allow for only discrete probabilities, in contrast to continuous probabilities.

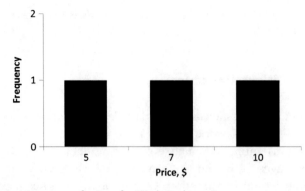

Fig. 2-1. Histogram of prices for Work Package A

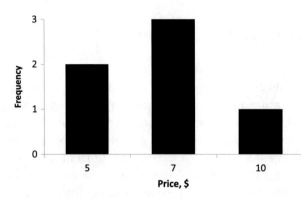

Fig. 2-2. Histogram of prices for Work Package B

2.6 Triangular Distribution

One approximation that greatly simplifies the analysis of the frequency distribution in Section 2.5 and allows for continuous analysis is the use of the triangular probability distribution as a surrogate for the normal distribution. It is possible to obtain a triangular distribution for optimistic, most likely, and pessimistic (OMP) prices by considering low-, medium-, and high-frequency levels (see Fig. 2-3). If an extreme price (the pessimistic price in this case) is expected to occur with a medium level of frequency, the triangular distribution could now result in a polyline, such as shown in Fig. 2-4. Thus, one triangle (A_1) and one nontriangular area (A_2) represent the distribution in those two ranges. Fundamentally, the triangular shape is the expected (or approximated) distribution profile for prices within the range of costs.

The varying frequencies of the prices could also be interpreted as the behavior of price data from different vendors if the data was requested from multiple vendors.

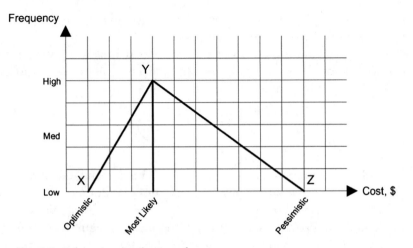

Fig. 2-3. Triangular distribution of costs

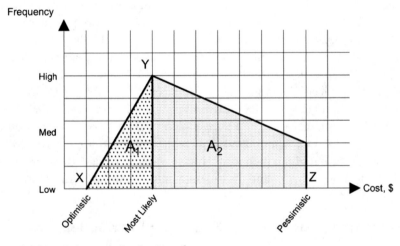

Fig. 2-4. Polygonal distribution of costs

2.7 Requesting Prices

Requesting prices for any particular work item is not necessary if the estimator has good intuition of the OMP (Optimistic—Most likely—Pessimistic) prices based on experience. That depends, however, on how the pricing is done. By combining that intuition with some feel for the probability of underrun or overrun, an experienced estimator can assign appropriate frequencies to the OMP prices and arrive at a triangular or polyline distribution. Thus, the estimator can determine the absolute minimum possible cost of a project—the optimistic price—as well as the absolute maximum possible cost—the pessimistic price—with other prices in between those extremes. The most likely price will, by definition, exhibit the highest frequency; otherwise it would not be the "most likely" price.

Whatever technique the estimator uses to request prices—whether based on market prices, historical records, or analysis from first principles—the bottom line is that each price must be divided into its OMP components.

2.8 Background

J. M. Neil (1982) first proposed the triangular distribution technique of risk analysis. The CII (1987) further advanced his approach, using a Monte Carlo method for replicating combinatorial work package costs for triangular distributions. However, for a small number of packages, the Monte Carlo method requires more time than Neil's method to make a similar determination because of its prolonged iterative process (as described in Chapter 3).

The Monte Carlo method of CII (1987) is redundant if the Monte Carlo is run thousands of times, as the report suggests, for a small number of work packages. For example, if the Monte Carlo ran thousands of times for 10 work packages, the result would be a combination distribution that reflects the original price distribution.

As opposed to the previously described CII method, the triangular distribution method enables a quick and simple project cost risk analysis for small projects or simplified large projects. For larger, more complex projects, a Monte Carlo simulation applied to the triangular distributions can be helpful and save time. Given, for example, a large number of work packages—say, 250,000—for a convention center project, half of which have considerable price deviation and are considered critical, the triangular distribution method would not fare well. That said, this chapter focuses only on the technique for quantitatively estimating risk without using Monte Carlo.

2.9 Probability Evaluation by Triangular Distribution

When using triangular distributions, it is assumed that the area under the curve represents the probability of the cost. Each work package comprises different elements of the project. For example, one work package may serve structural work, whereas another may serve mechanical work. A typical project has many (sometimes hundreds of thousands) work packages, and manual computations of all probabilities would be too tedious and uneconomical for a contractor when selecting a bid price. For this reason, on actual projects the use of computer software is recommended in order to obtain an accurate price. The technique that will be demonstrated is for an example project having only three work packages.

For purposes of discussion, let us say that an estimator determines three ranges of bid prices for three work packages of a particular project: Work Packages (WP) A, B, and C. Assume that the estimator's price data for Package A ranges from $4,000 to 8,000 with the most likely price at $5,000; Package B ranges from $2,000 to 5,000,

Table 2-1. Prices of Work Packages

Work Package	Price, $ (Thousands)		
	Lowest (Optimistic)	Most-Likely	High (Pessimistic)
A	4	5	8
B	2	4	5
C	8	10	12

with the most likely price at $4,000; and Package C ranges from $8,000 to 12,000, with the most likely price at $10,000, as shown in Table 2-1.

The goal is to determine some WP price profile representing the chances of occurrence of various prices in the continuum. Naturally, this will be a function of the probabilities of receiving low, most likely, or pessimistic prices on each work package.

Following data collection and organization, each work package needs to be analyzed individually. Fig. 2-5 depicts the triangular distributions for the three work packages. The areas under the curve represent the infinite number of possibilities for specific WP prices occurring at their respective frequencies. If an infinite number of bids were obtained, the distribution plot could be approximated to the triangular distribution graphs in Fig. 2-5.

To determine the probability of occurrence of any range of any work package, we first must compute the area under the curves. It is also known that the total area under the graph represents a 100% probability of occurrence. Thus, for Work Package A:

$$\text{Area of WP Triangle } A_1 = \frac{1}{2} \times 1 \times h = 0.5h \tag{2-1}$$

$$\text{Area of WP Triangle } A_2 = \frac{1}{2} \times 3 \times h = 1.5h \tag{2-2}$$

$$\text{Total Area of Work Package A} = \text{Area } A_1 + \text{Area } A_2 = 0.5h + 1.5h = 2.0h \tag{2-3}$$

where h is the height of the most likely frequency. The areas for other WP ranges are found similarly and are presented in Table 2-2.

Subsequently, the probability of occurrence of the zone represented by A_1 is

$$\text{Probability } (A_1) = \frac{\text{Area of } A_1}{\text{Total Area of WP A}} = \frac{0.5h}{2.0h} = 0.25 \tag{2-4}$$

Similarly, the probability of occurrence of the zone represented by A_2 is

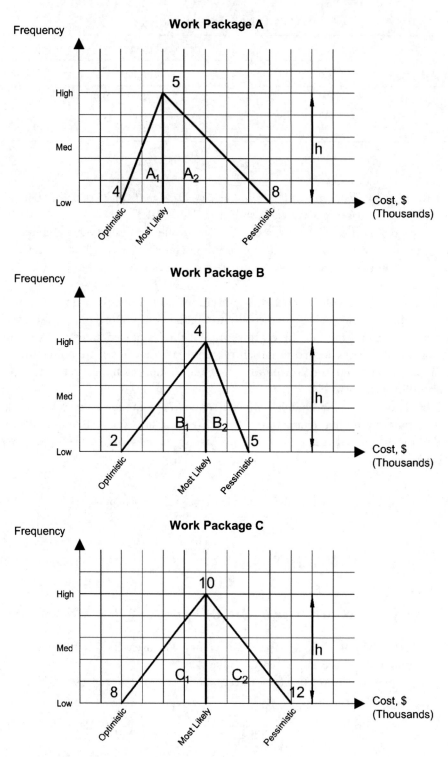

Fig. 2-5. Triangular distributions of costs for Work Packages A, B, and C

Table 2-2. Probability of Occurrence

Triangle	Area	Probability of Occurrence (P)
A_1	$0.5 \times (5-4) \times h = 0.5h$	$A_1/(A_1+A_2) = 0.5h/2.0h = 0.25$
A_2	$0.5 \times (8-5) \times h = 1.5h$	$A_2/(A_1+A_2) = 1.5h/2.0h = 0.75$
B_1	$0.5 \times (4-2) \times h = 1.0h$	$B_1/(B_1+B_2) = 1.0h/1.5h = 0.67$
B_2	$0.5 \times (5-4) \times h = 0.5h$	$B_2/(B_1+B_2) = 0.5h/1.5h = 0.33$
C_1	$0.5 \times (10-8) \times h = 1.0h$	$C_1/(C_1+C_2) = 1.0h/2.0h = 0.50$
C_2	$0.5 \times (12-10) \times h = 1.0h$	$C_2/(C_1+C_2) = 1.0h/2.0h = 0.50$

$$\text{Probability } (A_2) = \frac{\text{Area of } A_2}{\text{Total Area of WP A}} = \frac{1.5h}{2.0h} = 0.75 \qquad (2\text{-}5)$$

These are also presented in Table 2-2. One must check to see the probabilities of occurrence of $A_1 + A_2 = 1.0$, and they are.

2.10 Probabilities of Project Cost

To obtain the probability of the overall costs on the project and to determine the final bid price, the individual probabilities of occurrence for the work package triangles in Table 2-2 need to be combined. Joint probabilities for varying cost combinations are calculated by multiplying the individual probability of occurrence of each work package.

From Fig. 2-5, the dollar value represented by A_1 is somewhere between \$4,000 and \$5,000; B_1 is between \$2,000 and \$4,000; and C_1 is between \$8,000 and \$10,000. The midpoint of the range of each zone is taken to represent that zone's value. Thus, the occurrence of A_1 represents an incurred cost of \$4,500; for B_1, the cost would be \$3,000; and for C_1, it would be \$9,000. The occurrences of the corresponding subtriangles (i.e., $A_2, B_2,$ and C_2) can be obtained similarly.

From a mathematician's perspective, the centroid value could be used, because it is the "center" of the triangle. However, that would add extra calculations to what are already approximations, and the extra information provided might not be significantly more representative (in fact, it is restrictive because of the narrowing of the range of the considered spread). But to go deeper into this perspective of using the centroid as the representative value, the representative cost of the subtriangles in such case depends entirely on the frequencies of the triangular distribution. For instance, if the most likely cost has a very high frequency of occurrence, but the optimistic and pessimistic costs have very low frequencies of occurrence, the representative values in the subtriangles that share a border with the most likely cost will be close to the most likely cost. The representative costs in such cases will move away from the most likely cost as the frequency of the most likely cost diminishes and the frequencies of the optimistic and pessimistic costs increase. Note, however, that the frequency of the most likely cost will always be more than those of the optimistic or pessimistic costs; otherwise, the most likely cost would simply not be most likely.

When a discrete or specific point is considered (e.g., the midpoint) and that point represents a range of data, some information is lost. But, for expediency in estimating risks based on approximations, such loss of information is tolerable.

Table 2-3 shows the probability of occurrence and its corresponding cost—based on the midpoint—for each work package combination. The probability of occurrence is naturally the product of the probabilities of the individual work package zones, whereas the cost based on the midpoint value is the sum of the work package zones in combination. All eight possible combinations have been considered. Another way to determine how many combinations there will be in triangular cost estimation is to calculate m^n, where m is the number of split triangles of each work package (two each for WPs A, B, and C), and n is the number of work packages. Hence, this example yields $2^3 = 8$ possible combinations of prices.

From Table 2-3, it appears that the final price of *all* possible WP combinations ranges from \$16,500 to \$22,000. Arranging these work package combinations in ascending order aids in organization of the data (see Table 2-4). If there are multiple WP combination costs that are the same (e.g., $A_2 + B_1 + C_1$ and $A_1 + B_1 + C_2$), the probabilities of occurrence for all such WP combinations need to be combined. Table 2-4 shows how the probability of occurrence of WP combinations with the same cost is calculated and has the costs of these combinations arranged in ascending order.

2.11 Cumulative Probability of Project Costs

After calculating the joint probabilities and their midpoint prices, the cumulative probabilities were determined and are shown in Table 2-5. The price of \$18.5K

Table 2-3. Probability of Cost Occurrence for the Project

Combination of Work Package Zones	Cost Based on Midpoint ($, 1000s)	Probability of Occurrence
$A_1 + B_1 + C_1$	$4.5 + 3.0 + 9.0 = 16.5$	$0.25 \times 0.67 \times 0.50 = 0.08333$
$A_1 + B_1 + C_2$	$4.5 + 3.0 + 11.0 = 18.5$	$0.25 \times 0.67 \times 0.50 = 0.08333$
$A_1 + B_2 + C_1$	$4.5 + 3.5 + 9.0 = 17.0$	$0.25 \times 0.33 \times 0.50 = 0.04167$
$A_1 + B_2 + C_2$	$4.5 + 3.5 + 11.0 = 19.0$	$0.25 \times 0.33 \times 0.50 = 0.04167$
$A_2 + B_1 + C_1$	$6.5 + 3.0 + 9.0 = 18.5$	$0.75 \times 0.67 \times 0.50 = 0.25$
$A_2 + B_1 + C_2$	$6.5 + 3.0 + 11.0 = 20.5$	$0.75 \times 0.67 \times 0.50 = 0.25$
$A_2 + B_2 + C_1$	$6.5 + 4.5 + 9.0 = 20.0$	$0.75 \times 0.33 \times 0.50 = 0.125$
$A_2 + B_2 + C_2$	$6.5 + 4.5 + 11.0 = 22.0$	$0.75 \times 0.33 \times 0.50 = 0.125$
		Sum = 1.00

Table 2-4. Combined Probability of Cost Occurrence for the Project

Combination of Work Package Zones	Cost Based on Midpoint ($, 1000s)	Probability of Occurrence
$A_1 + B_1 + C_1$	16.5	0.08333
$A_1 + B_2 + C_1$	17	0.04167
$\left.\begin{array}{l} A_2 + B_1 + C_1 \\ A_1 + B_1 + C_2 \end{array}\right\}$	18.5	$0.25 + 0.08333 = 0.3333$
$A_1 + B_2 + C_2$	19	0.04167
$A_2 + B_2 + C_1$	20	0.125
$A_2 + B_1 + C_2$	20.5	0.25
$A_2 + B_2 + C_2$	22	0.125
		Sum = 1.00

Table 2-5. Probability of Project Cost Occurrence

Project Cost ($, 1000s)	Frequency of Occurrence	Joint Probabilities	Cumulative Probability
14.0*	–	~0.000	~0.000
16.5	1	0.0833	0.0833
17	1	0.0417	0.125
18.5	2	0.3333	0.4583
19	1	0.0417	0.5
20	1	0.125	0.625
20.5	1	0.25	0.875
22	1	0.125	1.000
25**	–	~0.000	1.000

*Minimum Cost.
**Maximum Cost.

occurs through two combinations, $A_2 + B_1 + C_1$ and $A_1 + B_1 + C_2$. Thus, the frequency of that cost is two. The frequencies of $16.5K, $17K, $19K, $20K, $20.5K, and $22K are each one. A graph of this project's cumulative probability profile can be plotted from this data (see Fig. 2-6). The cumulative probabilities will establish the most likely cost and the overrun and underrun probabilities.

2.12 Finding the Bid Price

The target cost, or target price, is defined as the total of the most likely cost of each work package. For the current project, this amounts to $A + $B + $C = $5K + $4K + $10K = $19,000. Fig. 2-7, which displays this target price, shows that it occurs at a confidence of 50%. Now, the questions to consider are

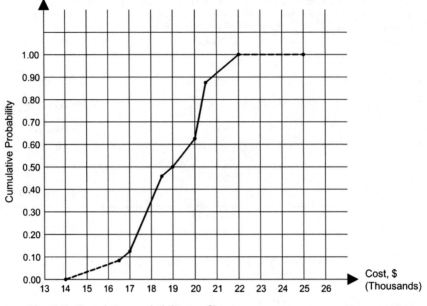

Fig. 2-6. Cumulative probability profile

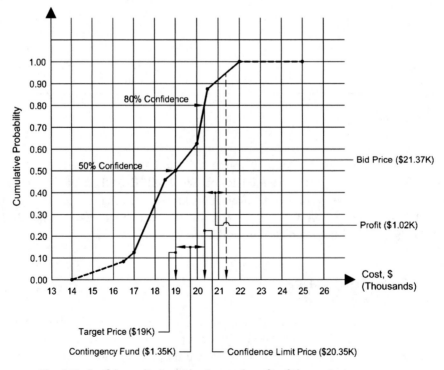

Fig. 2-7. Confidence limits, bid price, and profit of the project

- Would the contractor feel satisfied with bidding the work packages at this price when there is a 50% chance of overrun?
- Would the contractor be willing to take a risk in which the chance of overrunning the target price is equal to that of underrunning it?
- Is the contractor satisfied with these odds of 1:1?
- Should the contractor try to improve these odds, or should the contractor gravitate toward risk with the aim of winning the bid?

What should the contractor do? This seems to be a risk situation where *only* the contractor must make a decision. A reasonable decision involves ascertaining the confidence limit with which the contractor feels satisfied—that is, how much of a gamble the contractor is willing to take. A risk-averse contractor would bid with high confidence, whereas a contractor with less regard for risk might bid the target price or even lower.

Construction is a risky business; profits are rarely made without taking risks. However, many contractors incur losses because they take too many risks. Thus, choosing the final price of the work packages is a balanced trade-off decision.

Assume that the contractor feels safe with either the target price or the price at 80% confidence, whichever is higher. This is a rational heuristic in many cases. In this example, the 80% confidence level results in a price higher than the target price: $20,350. The odds of succeeding are now 4:1, much better than the previous 1:1 odds with just the target price. Thus, the contractor would aim to meet the target price, but would set aside a contingency fund equal to the confidence-limit price minus the target price (see Fig. 2-7) as seen in the following equation:

$$\text{Contingency fund} = \text{confidence-limit price} - \text{target price} \qquad (2\text{-}6)$$

In this example, Contingency fund = $20,350 − 19,000 = $1,350.

The bid price, however, is the confidence-limit price of the work packages plus profit. Thus, using this technique,

$$\text{Bid price} = \text{"confidence-limit" price of work packages} + \text{profit} \qquad (2\text{-}7)$$

or

$$\text{Bid price} = \text{target price} + \text{contingency fund} + \text{profit} \qquad (2\text{-}8)$$

If incorporating profit into the work packages themselves is considered, a problem arises because the profit from a work package is not subject to the same type of risk as the work package itself. Thus, different types of risks can get mixed up when trying to solve the problem in this manner. Either this type of construct of the work package should be avoided, or else the work package should be stripped of all profit when doing the risk analysis.

It is hoped that the project price will not increase beyond the confidence-limit price, but that is only a hope, and such an increase has a finite chance of occurring. In such a case, the contingency funds will be exhausted. Now assume that profit is 5% of the confidence-limit price. Recall that profit is a very personal item for the contractor but affects bid price.

Then, for the current case, from Eq. (2-7),

$$\text{Bid price} = \$20,350 + (0.05 \times \$20,350) = \$20,350 + \$1,020 = \$21,370 \qquad (2\text{-}9)$$

This bid price and profit value are shown in Fig. 2-7.

2.13 Additional Perspectives

2.13.1 The Owner's Perspective

The steepness of the *probability of project cost* curve will show how prices fluctuate between contractors. If the curve is steep, small fluctuations in price occur for that contractor; if the curve is flat, high fluctuations in price could occur. The owner should prefer a steeper curve, because the costs of a project will remain relatively consistent; therefore, other project factors apart from cost can be considered when choosing a contractor. Furthermore, claims can be expected to be less because the contractor is relatively unlikely to exceed the most pessimistic price for the project. However, the owner can use this graph only if it is requested from the contractor in the bid documents; owners have no way of making this graph themselves.

In addition, the owner can use this graph to determine how much contingency to obtain to finance the project. The contractors' bids typically include some contingency funding; contractors fund overruns that they feel could occur by bidding at a price greater than the most likely price. This bid price is determined by how much confidence the contractor has in underrunning the project. However, the owner should be inclined to consider additional contingencies, beyond the contractor's bid price, for any claims made by the contractor in excess of the original agreement owing to potential cost overruns. Owners can use the cumulative probability profile to determine—within a reasonable ballpark—their own level of confidence and thus the contingencies required to complete the project. That said, this is not the sole mechanism by which owners can determine project risk; they may use a wide variety of techniques ranging from competitive bid analysis (Singh and Shoura 1998) to market and inflation studies, among others.

2.13.2 Guaranteed Maximum Price (GMP)

A contractor can also use the cumulative probability profile to determine the guaranteed maximum price (GMP). For a GMP contract to occur, the owner and contractor must agree on a set price, including all contingencies within the bid price. In these contracts, the owner passes all risks to the contractor, and no claims beyond the GMP are entertained. Two scenarios can occur for the contractor under a GMP contract: Profit is made if the contractor spends less than the guaranteed maximum price; the contractor suffers a loss if spending exceeds the GMP. Because the owner and contractor agree on the GMP before the start of the project, the contractor may not make any additional claims against the owner.

The contractor can derive the GMP from the cumulative probability profile by assigning a reasonable and high confidence limit, with an associated price, that can cover most expected contingencies. A confidence limit of about 95% and a corresponding GMP can be expected in such an instance. However, it should be noted that this estimate is only possible if the technique outlined in this chapter is used to arrive at the cumulative probability profile. This technique enables the contractor to be satisfied as to the confidence of the GMP.

2.13.3 Contractor's Risk Transfer and Subcontractor's Risk

The most practical assurance for a contractor to minimize risk is to transfer risk to subcontractors. Contractors do so by accepting fixed prices from the subcontractor and holding the subcontractor to those prices if there is no scope change. When this happens, contractors have effectively locked their risks and have little need of probabilistic cost analysis. It is of the utmost importance for contractors not to interfere with the timetable or sequence of work proposed by the subcontractor, if this strategy is followed, and to refrain from interfering in any way that might disturb the subcontractors' plans and work schedule; otherwise, there can be legal consequences for the contractor. If the contractor accomplishes this, the risk imposed by the owner will successfully be transferred to the subcontractor. Large contracting companies can easily lessen their risk by transferring it in this manner, because transferring the entire contract to subcontractors may not be possible. Furthermore, the lowest level of subcontractors may not find it as easy to transfer their risk, because they are at the bottom rung of the contracting ladder and have no one to transfer their risk to. However, they can benefit from probabilistic cost estimation, although they may find it difficult to use depending on their educational levels and concept of usage of office time for cost analysis.

2.14 Summary

The triangular-distribution method serves as a simple way to evaluate the bid price and the probabilities of cost overrun and underrun for small projects or simplified large projects. The method uses triangular distributions to find the probability of getting a high, most likely, or low price from a work package. From this, the probabilities of combinations of various work packages that obtain specific total prices are determined. The total project prices with their probabilities of occurrence are arranged in ascending order, along with the calculation and plotting of the cumulative probability. The price for a satisfactory confidence of

Table 2-6. Project Costs for Problem 1

Work package	Price (thousands of dollars)		
	Lowest (optimistic)	Most likely	High (pessimistic)
A	25	29	36
B	17	24	28
C	14	21	35
D	30	36	40

risk is marked off the plot, and the contractor can thus allocate a known contingency fund to work packages for the degree of risk he or she feels comfortable with. Profit can be added over and above, and a final bid price can be determined. This technique can serve as a useful tool for owners to evaluate potential financial risks to themselves and discover maximum project cost. Its benefit to contractors is substantial.

2.15 Exercises

Problem 1

Given four work packages A, B, C and D, with their pessimistic, most likely, and optimistic prices given in Table 2-6, answer the following questions.

a. How many combinations of prices can we have with the four work packages given, using the triangular distribution method?
b. What is the most likely price? What is the price at 75% confidence? Present all your calculations and all tables and figures.
c. What would be the bid price, assuming 80% confidence and profit of 10%? Comment on the chances of the contractor winning this bid.

Problem 2

Repeat Problem 1, assuming that the optimistic prices are 10% lower and the pessimistic prices are 10% higher. What differences do you find from the results of Problem 1? Why?

2.16 Acknowledgments

This chapter is substantively modified from A. Singh, S. Shiramizu, and K. Gautam (2007), "Bid risk and contingency analysis." *Cost Eng.*, 49(12), 20–27.

References

CII (Construction Industry Institute). (1989). "Management of project risks and uncertainties." Publication 6-8, Austin, TX.

Molenaar, K. (2005). "Programmatic cost risk analysis for highway megaprojects." *J. Constr. Eng. Manage.*, 10.1061/(ASCE)0733-9364(2005)131:3(343), 343–353.

Neil, J. M. (1982). *Construction cost estimating for project control*, Prentice Hall, Englewood Cliffs, NJ.

Reilly, J., McBride, M., Sangrey, D., MacDonald, D., and Brown, J. (2004). "The development of a new cost-risk estimating process for transportation infrastructure projects." *Civ. Eng. Pract.*, **19**(1), 53–75.

Rubin, D. K. (2005). "Execs believe managing risk is their biggest challenge." *Eng. News Rec. Nov.*, **255**(21), 16.

Singh, A., and Shoura, M. M. (1998). "Optimization for bidder profitability and contractor selection." *Cost Eng.*, **40**(6), 31–41.

Touran, A. (2006). "Owners risk reduction techniques using a CM." CMAA Research Rep., Construction Management Association of America, Washington, DC.

Contingency Analysis and Allocation

3.1 Introduction: Contingency Cost Management

Once risk has been identified, a contingency fund can be allocated to mitigate risk items and unforeseen events to prevent potential losses, the aim being to measure and then control risks (Venkataraman and Pinto 2008). A contingency fund is a set amount of money placed aside in a project to help fund and cover potential losses [Construction Industry Institute (CII) 1989]. This advance risk-planning action, with its available funding, accepts that foreseen and unforeseen risks can occur. The amount of money placed in project contingency determines a contractor's ability to accept risks and mitigate potential losses. Should those risks not materialize, the contractor will come out on top, with those contingency funds added to profit. And should those risks materialize, the contractor will have contingency funds in the wings to address them. The determination and allocation of contingency funds is not perfect, but goes a long way to decide on contingency management by objective principles and mitigate surprises and uncertainties.

Before a project begins, it is difficult to fully predict actual project costs using only price estimates, because many events can intervene. These estimates may not include any unforeseen events, such as inflation, productivity loss, decrease in supply, material defects, new information that is brought to bear at the last minute, or new site findings—all of which would change the cost of project items. The contingency fund is set to budget for these unforeseen events and ensure that the project continues as smoothly as possible without any schedule delays. An added benefit of using contingency funds is that if items turn out to cost less than expected, the remaining funds set aside will become additional profit.

When estimates are made for each item cost based on past experience, various errors can creep in (Hajek 1977):

- Omissions: An estimate made can sometimes miss or forget to include necessary items for an accurate estimate.
- Misinterpretation: An estimate may be incorrect if the data provided were misinterpreted, leading to an estimate that is too high or too low.

- Failure to identify risk: Poorly identifying the project's risks may lead to inaccurate estimates.
- Inaccurate work breakdown: An inaccurate breakdown of each work package may lead to an inadequate amount of contingency funds allocated throughout the project timeline.

Hence, leaving budgets to past experience, without assessing the present situation and risks on the ground, can be a fatal mistake. Thus, a detailed analysis of specific risks is fundamental to ensuring that the project does not see red but makes, at minimum, the profit expected from it.

3.2 Contingency Cost Process Overview

This process is a guide for determining the amount to be set aside as a contingency fund. First, list the project's cost packages. Give each work package three estimated prices: a low price, a target price, and a high price. Such a triangulated price distribution is selected for its ease of practical application in the construction industry. The interested student and researcher can always opt for exotic probability distributions to simulate price distributions, but such sophisticated methods are rarely followed in the construction industry, for good reason. In addition, estimate the time expected to complete each work package in terms of percentage of time to total project completion. The purpose of this estimate is to allocate contingency funds according to a meaningful timetable; otherwise, the whole exercise could become quite theoretical and inapplicable.

To start, categorize all the cost items as critical or not, using a criticality rule. Then implement a Monte Carlo technique to assign the critical work packages to their respective probabilities. The Monte Carlo technique—although it takes its name from the casinos in Monte Carlo—here indicates that prices are generated at random to simulate what may happen in real life. Thereafter, possible scenarios are created either by a computer simulation, which can generate thousands of scenarios, or manually by random generation of numbers. The random generation of numbers is used to create a frequency distribution of expected costs from which a cumulative probability profile of expected costs can be extracted. From here, the contingency fund allocation can be set on the basis of the comfort level of the contractors in the bidding process.

An aggressive bidder will allow more risk in order to bid competitively, but this risk may result in a *project overrun*—the total project cost exceeding the bid price—resulting in a loss. A more conservative bidder will bid above the target price, aiming for a possible *cost underrun*—the total project cost being less than the bid price—resulting in a higher profit. Determine the amount of funds needed for the contingency pool based on how the possible cost overruns and project underruns

work out for each work package. Finally, draw a *drawdown curve* for the contingency fund to display the allocation of funds throughout the project's schedule.

3.2.1 Contingency Cost Allocation Methodology

Given here is an example of a cost contingency analysis in which the amount of contingency funds for a project is determined and a drawdown schedule prepared for allocating those funds. The estimate of each item's cost will include the item's low cost, target cost, and high cost, based on previous experience or recent quotes. These cost estimates are needed, because the cost items' prices may fluctuate once the project begins as a result of availability and demand. If supply is high and demand is low, then the item will gravitate toward the low cost; if supply is low and demand is high, then it will gravitate toward the high cost. However, if most variables remain the same, then the item will normalize around the target cost.

Next, assign a probability that each of the items' costs will occur. Although each cost item has already been estimated, the probability assigned will predict the amount eventually needed in the contingency fund as a result of potential cost overruns or underruns. The probability will help narrow the assumptions made in the estimates and guide us toward a better analysis of the contingency fund to be established. A safe but pessimistic way to determine the probability is to weight it more toward an overrun—that is, to expect higher costs. However, the drawback of this approach is that using higher estimated costs will diminish bid competitiveness. Conversely, management may deliberately minimize the probability of risks to keep prices down to make the bid more competitive.

Finally, assign a percentage of project completion to each cost item to help evaluate the project timeline and schedule once the item is completed. Afterward, the amount of contingency funds left and how they should be allocated throughout the rest of the project's schedule will be shown. Table 3-1 presents a sample case of a problem having four work packages: foundations, steel structure, civil works, and architectural works. The table shows low, target, and high prices, along with the probability of each. The last column on the right shows the percentage of the project work schedule accomplished when the corresponding work package is completed.

3.3 Critical Cost

3.3.1 Identifying Critical Work Packages

The first step in evaluating a work package for risk is to identify whether each cost of that work package is critical to the risk. There may be a long list of work packages, but the *critical* work packages are those whose price fluctuation varies from the target cost, either favorably or unfavorably, by greater than a set buffer value, or *criticality*

Table 3-1. Contingency Analysis Problem

Work package	Low price ($ millions)	Target price ($ millions)	High price ($ millions)	Probability of price occurrence	Project progress when work package is completed (% finished)
Foundations	5	6	8	Low = 0.1 Target = 0.3 High = 0.6	20
Steel structure	11	12	14	Low = 0.2 Target = 0.3 High = 0.5	45
Civil works	10	12	14	Low = 0.1 Target = 0.1 High = 0.8	95
Architecture works	2.99	3	3.01	Low = 0.4 Target = 0.4 High = 0.2	100
Total target price ($ millions)		33			

threshold. This is meaningful for prioritizing efforts to attend to the primary work packages that exert an influence on project price fluctuation.

In the contingency cost analysis for determining the critical work packages, a simple approach is to use the *criticality rule.* This rule states that an item is considered critical if its overrun or underrun price is greater than or equal to 0.5% of the project's total target price. Items costing less than 0.5% of the total target price may be considered noncritical, because varying their prices will not significantly alter the total target price. All noncritical cost items may be grouped together into one category held constant through the entire contingency cost analysis (CII 1989).

A typical technique for determining the criticality threshold is calculating whether the cost variation is within 0.5% (in conceptual cost estimates) or 0.2% (in detailed cost estimates) (Curran 1989). Other threshold values can be used at the project estimator's discretion. For example, if the conceptual target cost estimate of a project is $100,000, the criticality threshold for any work package, applying the 0.5% rule, is $500. A work package is considered critical when it is large enough to have a meaningful impact on the project's completion cost.

The critical work-package threshold focuses on the possible variation in the element's cost, not the actual cost. Continuing from the previous example, with the threshold at $500, a single element with a target cost of $50,000 will not be considered a critical work package if the product's overall cost does not have a variation of more than $500. Conversely, another element with a target cost of only $600 may be a critical work package if its price range varies from $400 to $1,200, because the maximum fluctuation ($1200 – $600 = $600) is greater than the $500 obtained by applying the 0.5% rule.

3.3.2 Handling Noncritical Work Packages

After the critical work packages have been identified and analyzed, the noncritical work packages will remain. They can be found by subtracting the critical work packages from the target project estimates. Within the Monte Carlo simulation, the noncritical work packages are grouped together as one fixed element. They have a high predictability, because their price variations for an overrun or underrun are very small. The noncriticals may have small fluctuations, but they are not enough to worry about in a project that has many other causes for worry.

3.3.3 Criticality Rule Examples

The total target price is calculated by adding up the target prices of all elements of the project. The total target price of this project is $33 million, with a variance for criticality of $0.165 million. The architecture works have a low estimated price difference of $0.01 million from the target and a high estimated price difference of also $0.01 million from the target, both less than the critical value, thus placing this item in the noncritical category. The rest of the items are placed under the critical item category, because each has a variance of greater than $0.165 million.

Moving on from the previous example, assume the project instead has the costs given in Table 3-2, to analyze a variation:

This project's total target price is $125 million, and the threshold variance for criticality is $0.625 million. As the table shows, civil works has a low price estimated at $0.5 million less than the target price and a high price estimated at $0.3 million higher than the target price. Both of these values are less than the critical value of $0.625 million, placing this work package, along with architecture works, into the noncritical category. The rest of the work packages are placed under the critical item category, as their variances are all greater than or equal to $0.625 million.

Table 3-2. Criticality Rule Example (Conceptual Cost Estimate)

Work package	Low price ($ millions)	Target price ($ millions)	High price ($ millions)	Maximum risk criticality ($)	Critical?
Foundations	34	35	38	3 (overrun)	Y
Steel structure	22	26	27	4 (underrun)	Y
Civil works	42.50	43	43.30	0.5 (underrun)	N
Architecture works	20.60	21	21.50	0.5 (overrun)	N
Total		125			

Table 3-3. Revised Cost Table

Work package	Low price ($ millions)	Target price ($ millions)	High price ($ millions)
Foundations	34	35	38
Steel structure	22	26	27
Noncritical items		64	
Total		125	

Therefore, Table 3-3 shows the cost table revised to lump civil works and architecture works into the single category of noncritical items.

3.4 Monte Carlo Simulation

Now we are ready to assemble the Monte Carlo *slabs*, which allow us to identify possible costs that may occur on the project. Monte Carlo slabs are ranges of random numbers (from 0.0 to 1.0) that represent specific dollar values for the work package. The range of random numbers depends on the probability of occurrence, as explained in Subsection 3.4.1. The multiple and repetitive generation of random costs enables a normalization of the possible price. Typically, simulations help increase the accuracy of the estimated prices of the risk elements, but these predictions will never be completely accurate, as the actual future is unknown.

A random number is assigned to each low value, target value, and high value on the basis of the probability that the value will occur. The noncriticals are not assigned a random number, because they do not vary. The generation of the Monte Carlo slabs follows directly from the probabilities of the optimistic price, most likely price, and pessimistic price. Hence, a technique of either judgment based on experience or of discovering the area under the price distribution—be it triangular or otherwise—is necessary to determine the probabilities of price occurrence.

Using a computer simulation to generate thousands of scenarios would serve as a more accurate technique, because it allows more cases of price occurrence to be considered. However, for simplicity's sake, this example will only use random generation for a small sample of 12 scenarios. Depending on the random numbers generated, we will determine the price of each work package and the price of the project.

3.4.1 Creating the Monte Carlo Slab

The Monte Carlo slab is created based on the probabilities of getting a low, high, or most likely price. For the foundation work package of Table 3.1, for example, the first Monte Carlo slab ranges from 0 to 0.1, because the chance of getting a low price for the foundation is only 10%. The probability of getting the target price is 30%, so we add 0.3 to 0.1 to get 0.4 for the upper value of the slab. Hence, the second slab

Table 3-4. Monte Carlo Setup

Work package	Value ($ millions)	Random number Monte Carlo Slab
Foundations	5	0.0–0.1
	6	0.1–0.4
	8	0.4–1.0
Steel structure	11	0.0–0.2
	12	0.2–0.5
	14	0.5–1.0
Civil works	10	0.0–0.1
	12	0.1–0.2
	14	0.2–1.0
Noncritical items	3	

ranges from 0.1—where the low price drops off—to 0.4. The probability of getting a high price is 60%, so we add 0.6 to 0.4 to get 1.0; the last slab will then range from 0.4 to 1.0. In essence, the Monte Carlo slabs for each element must span a range of 0.0 to 1.0, because all 100% of the possible occurrences must be accounted for. Table 3-4 illustrates the Monte Carlo slabs created for all the work packages.

3.4.2 Generating Random Numbers

At this stage, a random number is generated using a simple scientific calculator such as all engineering students have. This random number must fall between 0 and 1.0, because the probability cannot exceed 1.0 nor be less than 0.0. The random number that is obtained dictates the price of the element we will accept for the purpose of this simulation. So if we obtain a random number value of 0.5504 for foundations, it falls between 0.4 and 1.0; because that range is associated with a price of $8 million, the Monte Carlo slab assigns that $8 million value to the work package. This is the simple mechanism for creating and using a Monte Carlo slab. In like manner, we generate random numbers for all critical work packages and assign prices. These prices are listed in Table 3-5, based on the 12 simulation runs made for each work package. The reader can figure out what random numbers were probably generated for each run and each work package.

It is better to generate an independent random number for each work package at each time, because doing so represents independence and greater randomness, which is the more realistic scenario in the real world. Conversely, using the same random number to generate prices for all work packages would indicate that the price occurrences of the work packages are inter-connected and dependent on each other. The choice of generating a unique random number for each work package or one for all work packages is the decision of the estimator.

Table 3-5. Random Number Generation

Work package	Run (all prices in $ millions)					
	I	II	III	IV	V	VI
Foundations	8	5	6	8	6	8
Steel structures	14	11	12	14	12	14
Civil works	13	10	14	14	14	14
Total	35	26	32	36	32	36
	VII	VIII	IX	X	XI	XII
Foundations	8	6	6	5	8	6
Steel structures	14	12	12	11	14	12
Civil works	14	14	14	10	14	12
Total	36	32	32	26	36	30

3.5 Cumulative Probability Profile

The total price of the critical elements is seen to range from $26 million to $36 million in the 12 runs.[1] Now we will create a frequency distribution. The price of $26 million occurs *two* times, the price of $30 million occurs *one* time, and so on. Out of the 12 occurrences, $26 million occurs $2/12 = 1/6 = 16.67\%$ of the time; $30 million occurs $1/12$ of the time; and so forth. The probabilities of occurrence can thus be arranged in ascending order of price and a cumulative probability calculated. The sum of the probabilities and the cumulative probability must add up to 1.0, because 100% of the price distribution curve must stand covered. The results in Table 3-6 were used to create the cumulative probability profile shown in Fig. 3-1. For convenience, and also because additional detailed information is unknown, we assume that straight lines connect the various points in the figure. The reader can plan on drawing a smooth curve, if desired, but without an equation for that curve, the values on the y-axis for any price will have to be discovered by scaling and drawing. Based on the profile, if the target price is set at a low probability of confidence, it means that the project will have a greater chance of winning the bid but will more likely overrun and thus require a larger contingency fund to bid safely on the project. If the target price is set with a higher probability of confidence, it means that the project will have a lower chance of winning the bid, but the bid will more likely underrun, in which case a smaller contingency fund is needed to ensure a safe bid.

The target price of this project is $33 million. The cumulative probability profile shows the probability of overrun and underrun from the target price, as shown in Fig. 3-1. In this figure, the project underruns are seen below the target price point and overruns above it. In this example, the target price of $33 million has a probability of

---

[1]The Monte Carlo generates these results even as it is noted that the lowest possible cost of the project is $28.99 million, while the highest possible cost is $39.01 million.

Table 3-6. Frequency Distribution

Value ($ millions)	Non-critical ($ millions)	Total price ($ millions)	Frequency count	Probability	Cumulative probability
26	3	29	2	1/6	1/6
30	3	33	1	1/12	1/4
32	3	35	4	1/3	7/12
35	3	38	1	1/12	2/3
36	3	39	4	1/3	1

Note: 80% cumulative probability, or 80% confidence, will be represented by $35.4 million.

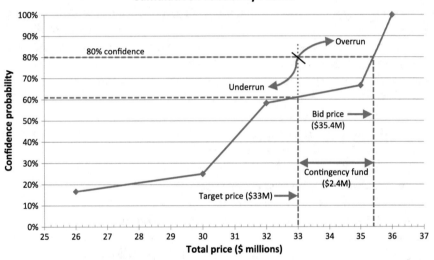

Fig. 3-1. Cumulative probability profile

about 61%, which indicates a low probability of an underrun if the project is bid at the target price; correspondingly, the project has a 39% chance of overrunning the target price, which could be bothersome to the company. In the latter case, to ensure the project will not overrun, a large contingency fund should be allocated.

3.6 Contingency Allocation

A safe bid with a 100% confidence level will eliminate the risks of an overrun, should there be one, while ensuring profit but will result in a tremendously low probability of winning the bid in a competitive environment. The lowest price a bidder should offer to increase the chance of winning the bid depends on the bidder's personal confidence level, which is often a gamble. But for the bidder to develop an appropriate confidence level is challenging indeed. Many a contractor has been successful by arriving at a reasonable confidence level, whereas many another has failed altogether by using poor judgment. Bidders' confidence levels vary, depending on their individual appetite for or aversion to risk. A good bidder will be able to bid competitively for a project at the proper price without leaving money on the table and while covering contingencies for possible cost overruns.

This example assumes that the bidder prefers to operate at an 80% confidence level in making the bid, which permits bidding very conservatively, because the target price sits below 61% confidence. An 80% confidence level translates to a bid price (before profit) of $35.4 million, with the target price of the project at $33 million. This means that $35.4 million – $33 million = $2.4 million will be put into the contingency fund. *This is the total allocation for contingency costs.*

The reason that the contingency fund is not $3.8 million[2]—the total of the expected overruns on the work packages—is that only 80% confidence was used in deciding the project cost (without profit). The 80% confidence value follows the profile of the cumulative probability, which is in no way fully linear across the span of the profile.

To allocate funds among the work packages, calculate the difference between the expected overrun and the expected underrun. Contingency funds will be assigned to that work package only if there is a net expected overrun, for obvious reasons. In the example, foundation works have a net overrun of $1.1 million, steel structures $0.8 million, and civil works $1.4 million. Hence, all three will be considered in the cost allocation. Noncriticals will not be considered at all, because they don't need any risk allocation. Table 3-7 shows this detail. Note that the probabilities of overrun and underrun are assigned from past experience and variable market conditions.

Next, the contingency funds are allocated, based on the ratio of their net overrun to the total expected overrun, $3.3 million. Because the total contingency allocation is $2.4 million, it follows that foundation works get allocated $0.80 million, steel structures $0.58182 million, and civil works $1.01818 million.

To this point, we have allocated the contingency funds based on the perceived risk of expected overrun on the work packages. But questions remain: How do we monitor expenditures? When do we expend the contingency funds? And what timetable do we follow to draw on these funds? These are important questions, because not only does a timetable alert project managers when they can expect to draw on the contingency funds, but it also allows cost controllers to obtain investment interest on those funds while they are waiting to be drawn. The next section thus looks at the drawdown timetable.

3.7 Drawdown Curve

Table 3-7 shows the allocation amount and also repeats the work package percentage completion that was given in Table 3-1. The drawdown curve simply connects the

---

[2]$2 × 0.6 + $2 × 0.5 + $2 × 0.8 = $3.8 million.

Table 3-7. Contingency Allocation Table

Work package	Overrun ($ millions) (A)	Underrun ($ millions) (B)	Probability of underrun (C)	Probability of overrun (D)	Expected overrun-underrun ($ millions) (E)	Allocation (%) (F)	Contingency fund allocation = (F) × total contingency ($ millions) (G)	Completion (%) (H)
Foundations	2	1	0.1	0.6	$(2 \times 0.6) - (1 \times 0.1)$ = 1.1	1.1/3.3 = 33.33	0.3333×2.4 = 0.80	20
Steel structure	2	1	0.2	0.5	$(2 \times 0.5) - (1 \times 0.2)$ = 0.8	24.24	0.2424×2.4 = 0.58182	45
Civil works	2	2	0.1	0.8	$(2 \times 0.8) - (2 \times 0.1)$ = 1.4	42.42	0.4242×2.4 = 1.01818	95
Noncriticals	–	–	–	–		–	–	100
Total					3.30	100	2.40	

Note: Total contingency = $2.40 million.

contingency fund to the time by which the work package is planned to be finished per the project schedule. Inasmuch as the project schedule is an estimate, if it is inaccurate, the drawdown curve will be inaccurate accordingly. But it is still better to have an estimate of the drawdown than no idea at all.

We started with a contingency fund of $2.4 million. This represents 100% of the contingency fund. As we approach the end of the drawdown, we will likely be left with contingency funds that approach zero, depending on how the funds are used. Because foundation works, steel structure, and civil works are the only work packages being allocated contingency funds, we see that the contingency funds for foundation works ($0.80 million) can be expected to be entirely consumed by the time 20% of the project is finished. By the time structural works are finished, at 45% completion, the contingency funds for steel structure ($0.58182 million) are expected to be consumed as well. Hence, at 45% project completion, the contingency funds for foundation works and structural works have both been consumed.

The drawdown curve is completed in this fashion until all work packages carrying contingency funds have been completed. Hence, 100% of the contingency fund allocation, $2.40 million, has been expended when civil works are completed, at 95% project completion. The drawdown curve thus derived is given in Fig. 3-2, which shows both percentage of funds depleted and dollar value of funds in contingency versus project completion.

It can be seen that there are three turning points on this project: one at 20% completion, another at 45% completion, and the third at 95% completion, when the contingency funds are expected to have been depleted and the drawdown curve comes to a stop on the x-axis.

If funds are left over after the first two work packages are completed, they may still be used for other items in the project in case of overrun. If funds are still left over

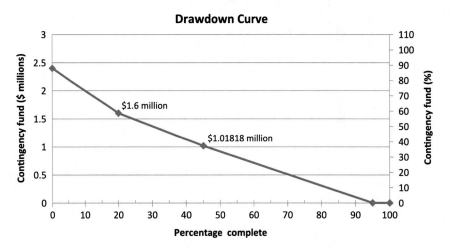

Fig. 3-2. Drawdown curve: contingency fund vs. completion time

once the entire schedule has been completed, they can be seen as a windfall—an extra profit from smart bidding.

3.8 Summary

This chapter has presented the quantitative basics of risk in contingency-cost analysis. In any engineering project, the proper steps are essential in order to manage risk so that the project can be a success. A contingency fund can insure against potential losses on the project and may benefit the project with added money left over after completion. A thorough contingency-cost analysis includes the following steps:

1. Lay out the project. Identify reasonable estimates for each work package at a low cost, target cost, and high cost. Assign probabilities to each of the probable costs. Indicate an estimated project completion percentage for each cost item.
2. Identify critical and noncritical cost items. A simple check to use is the 0.5% criticality rule for conceptual cost estimates, though other criteria can be developed.
3. Set up a Monte Carlo simulation to assign probabilities to cost items for possible case scenarios.
4. Use the Monte Carlo slabs to simulate possible outcomes. A simple random generation of numbers can be used for a smaller sample size.
5. Sort the simulations into a frequency distribution to find the probability of each possible scenario.
6. Use the frequency distribution to draw a cumulative probability profile.
7. Based on the cumulative probability profile, decide the proper amount to be allocated to the contingency fund. An aggressive bidder will bid much lower, with a lower confidence level, and place a more competitive bid, whereas a conservative bidder will bid higher, with a high confidence level.
8. Generate a drawdown curve to determine how the contingency fund will be allocated throughout the project's schedule.

   By following the proper steps in the contingency fund analysis, one can place a suitable allocation in the contingency fund, whether an aggressive or a conservative bidder.

3.9 Exercises

Problem 3-1

Assume that a project contains the costs given in Table 3-8. Determine which costs are critical and noncritical based on the criticality rule.

Table 3-8. Costs for Problem 3-1

Work package	Low price ($ millions)	Target price ($ millions)	High price ($ millions)
Utilities	5	6	7
Concrete structure	1	2	3
Asphalt	3.99	4	4.05
Steel	7.90	8	8.10

Problem 3-2

Table 3-9 gives the frequency distribution for a project. The target price of the project is estimated at $18 million. Draw the cumulative probability profile to determine which costs will be overrun and which underrun.

Problem 3-3

Table 3-10 gives the cost and completion data for a project, along with the expected probabilities of overrun. Run the Monte Carlo simulation and prepare a complete allocation schedule at a confidence level of 75%. Assume any additional data necessary.

Table 3-9. Frequency Distribution of Costs for Problem 3-2

Project price ($ millions)	Non-critical costs ($ millions)	Total price ($ millions)	Frequency count	Probability	Cumulative probability
13	5	18	2	1/5	1/5
14	5	19	2	1/5	2/5
18	5	23	2	1/5	3/5
22	5	27	4	2/5	1

Table 3-10. Costs for Problem 3-3

Work package	Low price ($ millions)	Target price ($ millions)	High price ($ millions)	Probability of overrun (%)	Completion (%)
A	20	25	$35	66	20
B	15	25	$30	33	40
C	10	12	$17	70	60
D	9	14	$18	45	90
Noncriticals	8	8	$8		

References

Construction Industry Institute (CII). (1989). "Management of project risks and uncertainties." Publication 6-8, Austin, TX.

Curran, M. W. (1989). "Range estimating: Measuring uncertainty and reasoning with risk." *Cost Eng.*, **31**(3), 18–26.

Hajek, V. G. (1977). *Management of engineering projects*, McGraw-Hill, New York.

Venkataraman, R. R., and Pinto, J. K. (2008). *Cost and value management in projects*, Wiley, Hoboken, NJ.

CHAPTER 4

Cause-and-Effect Diagrams

4.1 Introduction

This chapter examines the important quality management tool of *cause-and-effect diagrams* (CEDs), also known as *Ishikawa diagrams* or *fishbone diagrams*. The CED is a simple concept that allows groups to determine the root causes of outcomes in systems, processes, and operations. There are three different types of CEDs: dispersion analysis, production process classification, and cause enumeration. This chapter discusses CEDs through applications, examples, and practice problems, and also explains their construction.

The CED is one of the original instruments for production quality management. CEDs are most often used to improve production quality, but they can be broadly applied to virtually any field of study and can help improve the level of understanding of problems or the quality of outcomes in any system, be it environmental, biological, or social. As the first step in problem-solving, they organize a team's brainstorming efforts, which allows team members to think of ideas they would not have thought of individually. The CED serves as a systematic tool to determine all possible causes of a problem or outcome and to organize similar causes into groups to enhance the understanding of risk and minimizing it. Ideally, the CED diagnoses the root cause and focuses the team's analysis of a problem. The causes are arranged by importance in a ladder of causes and can then be related to each other. Fig. 4-1 illustrates a basic CED where it can be observed how various factors (causes) lead to outcomes (effects). The same concept can be presented differently as in Fig. 4-2, where various influencing categories, subcategories, and basic categories (all causes) lead to a specific problem or issue (effect) that needs to be analyzed. By simply drawing the CED, causes can be identified, isolated, and then scrutinized to determine what is causing the problem we wish to solve. Risks of various types are such effects that can be mitigated by understanding and alleviating their causes through such a structured exercise.

4.2 Kaoru Ishikawa

Dr. Kaoru Ishikawa, one of the founding fathers of modern management, viewed quality improvement as a continuous process. In 1943, Ishikawa developed the first

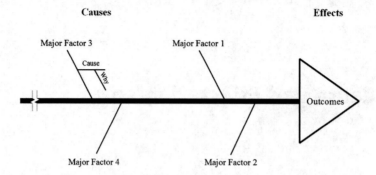

Fig. 4-1. Basic cause-and-effect diagram (modified from Simon 2010)

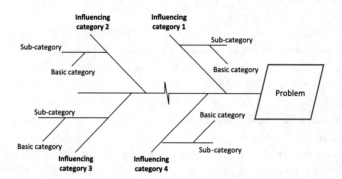

Fig. 4-2. CED with influencing categories, subcategories, and basic categories that can be analyzed

CED—hence the name *Ishikawa diagrams*—at the University of Tokyo. CEDs are also referred to as *fishbone diagrams* because of the similarity between the diagram's shape and a side view of a fish skeleton (Ishikawa 1976). Ishikawa used CEDs to demonstrate to engineers at the Kawasaki Steel Works that causes are interrelated.

Ishikawa also contributed to the growth of the concept of quality circles and published several books, including *Guide to Quality Control* (1976). Topics discussed included histograms, cause-and-effect diagrams, check sheets, Pareto diagrams, graphs, control charts, scatter diagrams, binomial probability paper, and sampling. Ishikawa is credited with helping Japanese industries rebound after World War II by contributing to quality improvement. He was held in such high regard, and his contributions were considered so very important, that in 1993 the American Society for Quality established the Ishikawa Medal to honor leadership in improving the human aspects of quality (American Society for Quality 2010).

4.3 Applications of Cause-and-Effect Diagrams

According to Ishikawa (1976), CEDs are most successful when brainstorming with a group. It is ideal to have as many people as possible contribute ideas. Although some would expect such a large group to cause disruption, participants can brainstorm a larger spread of ideas that can be narrowed later, whereas if one starts with a narrow spread of ideas, it can be quite difficult to widen it later.

Simple CEDs without a lot of second- or third-degree causes usually indicate incomplete knowledge of the subject and are considered inadequate because of their lack of important detail. Conversely, going beyond fourth-degree causes can make the CED too complicated and academic and can detract from its purpose. Hence, only a reasonable depth of detail, fit for the purpose, makes sense in a construction or production setting.

CEDs serve as a useful communication tool when working with both quantifiable and unquantifiable data. Fig. 4-3 illustrates the different notations used in CEDs (Ishikawa 1976). Underlined causes indicate unquantifiable causes with definite relationships to the outcome. A circled date next to a cause indicates that the cause was tested and determined to directly cause an outcome on that calendar date. Causes with quantifiable data should be noted in a box. Quantifiable data should then be collected and analyzed with histograms. The mean and standard deviation of the collected data should be obtained and then incorporated into the diagram.

CEDs are also applied in project management. Companies use different types of CEDs to visualize the connections among different causes of a problem. For example, a *production process classification* CED can help companies determine how various factors interrelate and affect a sequence of events. CEDs are used as part of the Six Sigma business-management strategy. Many companies, including Toyota Motor Corporation, use Deming's system of profound knowledge and CEDs in their operations to develop their management strategy. This has resulted in improved quality control and fewer communication lapses. CEDs are also used in product design. For instance,

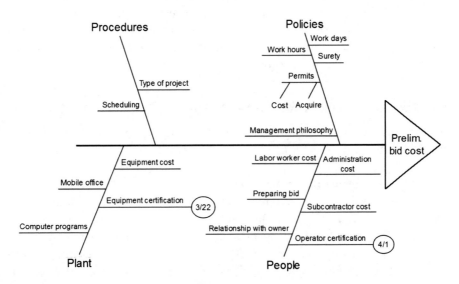

Fig. 4-3. Example cause-and-effect diagram (CED)

when Mazda Motors designed the Miata sports car, major causes in the fishbone diagram included touch and braking; minor causes included factors such as 50:50 weight distribution and the ability of the driver to rest an elbow on the door. The final production design included all the factors raised in the CED (Goodden 2009).

In effect, the CED helps to narrow uncertainty as well as to educate the engineers and workers on how best to manufacture or construct the parts for which they are responsible. Together, they have a synergistic effect on quality and success.

4.4 Types of Cause-and-Effect Diagrams

Ishikawa (1976) describes three types of CEDs:

- *Dispersion analysis* is the most popular type. "Why does this dispersion (outcome) occur?" is the key question used to construct a dispersion diagram. Dispersion analysis organizes and shows the relationships between causes and effects.
- The *production process classification* type examines the outcome in stages in a sequential process that follows the fishbone spine. Thus, the CED bones represent stages in the process and can be depicted as an assembly line. This type of CED is easy to understand but causes can be repetitive throughout the diagram. The production process classification CED is not confined to the production industry but can also be used for other outcomes (Omachonu and Ross 2004).
- The *cause enumeration* CED is similar to the dispersion analysis type except that it is constructed by a different method. The group brainstorms and lists all possible causes of an outcome before constructing any part of the CED. It is then constructed to relate the list of causes. The cause enumeration type of CED is more comprehensive because of the uninterrupted brainstorming. However, constructing a CED this way will possibly be more difficult because it is harder to relate the causes to one another (Ishikawa 1976).

4.4.1 Drawing the Dispersion Analysis Type of Cause-and-Effect Diagram

The following paragraphs summarize the different steps involved (Brassard and Ritter 1994) in constructing a dispersion analysis CED.

1. **Identify the outcome (dispersion):** The group can choose any problem or event to analyze for the outcome statement. The outcome should be positioned in a box on the right-hand side—in other words, the effect side—of the CED. The outcome box represents the head of the fishbone. Sufficient writing space is necessary to properly construct a CED, because group members must concur on the outcome statement and should include as much information as possible. Fig. 4-4 gives an example of an outcome statement.

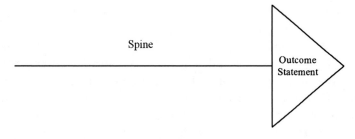

Fig. 4-4. Dispersion analysis CED, Steps 1 and 2

2. **Draw the fishbone spine:** The spine should extend horizontally to the left from the effect box. The spine connects all causes to the cause side of the CED and the cause side to the effect side. Fig. 4-4 illustrates the spine of a CED.
3. **Identify major-cause categories:** The major-cause categories are the bones of the fish, in other words, diagonal lines attached to the fishbone spine. Standard major categories used in the service industry are the four Ps: policies, procedures, people, and plant (technology) (Simon 2010). Fig. 4-5 illustrates a sample of a service industry cause-and-effect diagram. Standard categories used in the manufacturing and production industries are the six Ms: machines, methods, materials, measurements, Mother Nature (environment), and manpower (people) (Goodden 1951). In construction, one could use materials, labor, equipment, and subcontractors (MLES). The standard categories are guidelines and can be modified for any given problem. There is no fixed number of major-cause categories that must be used in the CED.

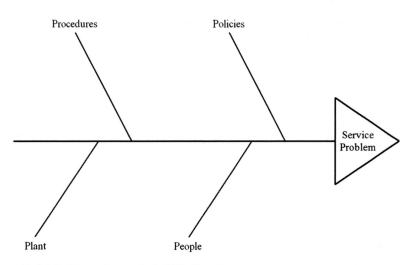

Fig. 4-5. Dispersion analysis CED, Step 3

4. **Brainstorm and categorize other causes:** Identify all causes that contribute to the outcome, and place each cause in the appropriate major category. The secondary causes are smaller bones placed on the major bone. A cause should be placed in all applicable categories, even if that means repeating it. Group members should develop a system to ensure that each member contributes equally to the discussion. One possible brainstorming system is having participants take turns, with each contributing one cause per turn. Fig. 4-6 illustrates this step.

5. **Further scrutinize each cause and study its reason:** Third-degree causes are bones attached to a secondary-cause bone (Fig. 4-7). Theoretically, the CED could continue to grow forever to infinity, because everything derives from a cause.

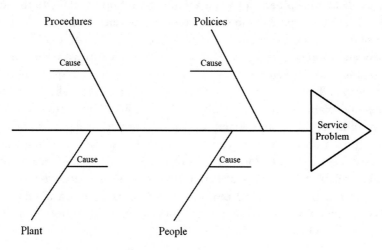

Fig. 4-6. Dispersion analysis CED, Step 4

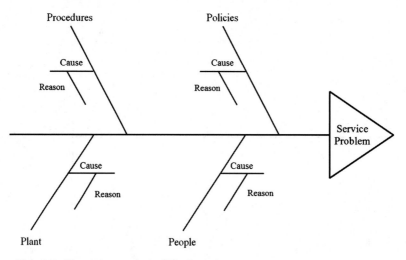

Fig. 4-7. Dispersion analysis CED, Step 5

The group must decide when to stop the CED's growth. Goodden (1951) suggests stopping it when causes become more than one level of management removed from the group analyzing the causes. At other times, going two to three levels deep may suffice. A very practical approach has to be adopted in this regard.

6. **Analyze the CED and determine which causes should be individually studied:** Repetitive causes might indicate a root cause. The diagram should be examined for clusters of causes, which may lead to further study. Quantifiable causes should be measured because they can help gauge changes to the outcome. A Pareto chart[1] may serve as a helpful tool to determine a possible root cause.[2]

Example 1 demonstrates a dispersion analysis type of CED.

4.4.2 Drawing the Cause Enumeration Type of Cause-and-Effect Diagram

The *cause enumeration* type of CED results in the same diagram as the dispersion analysis type. The difference between cause enumeration and dispersion analysis is the procedure involved when drawing the CED.

The first step in constructing a cause enumeration CED is to determine the problem or outcome to be studied. The next step is to brainstorm a list of causes for the output. After making a complete list of causes, main cause categories must be determined; each cause is then placed accordingly on the diagram. Further brainstorming for third-degree causes can be done after a preliminary CED is laid out. The finished cause enumeration CED should be identical to a finished dispersion analysis type. Finally, group members must analyze the CED to determine the root causes of the problem or output. Fig. 4-7 shows a finished cause enumeration CED.

Example 3 demonstrates a cause enumeration type of CED.

---

[1]A Pareto chart is common in quality control, where it highlights the most important factor among a set of multiple contributing factors, ranked in descending order of importance. Consequently, a Pareto chart enables a quick discovery of the most frequent reasons for defects, complaints, or other matters being tracked.

[2]The Pareto chart is named after Vilfredo Pareto, 1848–1923. Interestingly, Pareto was trained as a civil engineer, earning his doctorate with a thesis titled *The Fundamental Principles of Equilibrium in Solid Bodies*. He worked seven years with the Italian Railway Company and the private industry and then was general manager of Italian Iron Works. From civil engineer, he transformed to an economist when in his forties, and later in life he wrote a treatise on social science titled *Mind and Society*, which is his best-known work. He served as chair of political economy at the University of Lausanne, later making his most famous observation that 80% of the people in Italy owned only 20% of the land; this was generalized by quality guru Joseph Juran as the 80:20 rule, which he referred to as the Pareto principle. (Ref. Amoroso, Luigi (January 1938). "Vilfredo Pareto". *Econometrica.* **6** (1): 1–21. doi:10.2307/1910081. JSTOR 1910081; Aspers, Patrik (April 2001). "Crossing the Boundary of Economics and Sociology: The Case of Vilfredo Pareto". *The American Journal of Economics and Sociology.* **60** (2): 519–45. doi:10.1111/1536-7150.00073. JSTOR 3487932.

4.4.3 Drawing the Production Process Classification Type of Cause-and-Effect Diagram

The method behind the *production process classification* type of CED is different from that for the dispersion analysis and cause enumeration types. Group members should search for the reason behind each cause through, for example, dispersion analysis. Each process has an outcome and consequent causes of the outcome (Ishikawa 1976). The production process classification CED is similar to an assembly-line process: one keeps adding causes, step by step, in a specified order as the product completion advances.

The first step in the production process classification CED is to identify the problem or outcome to be studied. After that, the group should determine the process behind the outcome and how the outcome is achieved. Major-cause categories should serve as milestones in the outcome process; once they are determined, group members should brainstorm to find second- and third-degree causes. Finally, group members should analyze the CED to determine the root cause of the outcome. Figs. 4-8 and 4-9 represent possible production process classification types of CEDs. Fig. 4-9 is similar to an assembly line.

It must be noted that there are only minor shades of difference among the three types of CEDs. The major lesson to take away is that all types of CEDs hunt for causes and the reasons for those causes.

4.5 Example Problems

Three examples of the different types of CEDs demonstrate the various methods.

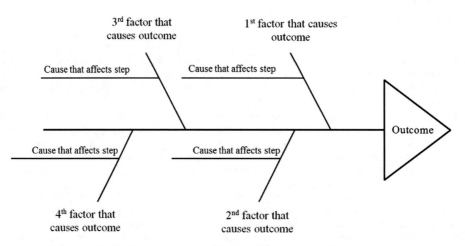

Fig. 4-8. Production process classification CED, example A

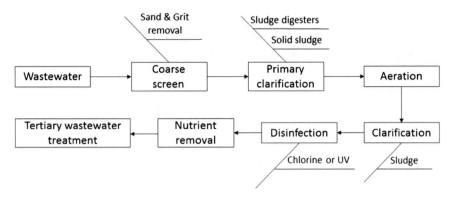

Fig. 4-9. Production process classification CED, example B

4.5.1 Example 1: Dispersion Analysis Type

Let's construct a CED using an example that readers of this book can relate to.

The U.S. Department of Education (2009) reported that the majority of first-time college students complete their bachelor's degree within six years. To examine the reasons that students do not graduate within four years, an investigatory team with three members decided to use a dispersion analysis CED for their study. The major-cause categories included policies, other obligations, other people, and the students themselves.

The major-cause category of *policies* was taken up for deeper analysis. Secondary causes categorized in policies that affected students' graduation time included the minimum grade requirement of C or better, course prerequisites, major requirements, the A to F bell-curve grading system, the price of education, and limited class sizes. The C-or-better policy prevents students from progressing in their studies and taking higher-level courses if they do not receive the requisite grades. Course prerequisites limit a student's flexibility in class choices and the number of classes the student can take each semester. Not fulfilling major requirements prevents students from receiving admission to their desired college and taking major classes. The A to F bell curve grading policy will always fail a few students, meaning that it will delay graduation. Rising tuition costs will increase the price of education and affect whether a student can afford to continue in college. Third-degree causes for the price of education include inefficient budgets and the economy. Limited class sizes can further reduce students' chances of getting their desired courses, which can delay their graduation; at the next level, limited classroom space and teacher preferences can limit class sizes, which also affects graduation time.

The major-cause category of *other obligations* was then examined. The secondary-cause categories that the investigatory team felt affected students' graduation time included working, being family providers, religion, and military service. Working directly affects the amount of time students can devote to their studies. Third-degree causes for students' obligation to work could be the need to support themselves and dependents and to pay for education. Being a family provider—acting as financial support for siblings, parents, or children—can cause additional stress and affect the amount of time a student can devote to studies. Students' faith or religion can affect their studies; for example, a student might need to go on a religious mission to fulfill the obligations of his or her faith. A student in military service could be deployed to war or training, thus interrupting study time. Yet students may join the military out of patriotism, to fund their education, or to obtain other financial benefits.

The major-cause category *other people* was then examined. Second-degree cause categories that the investigatory team felt affected a student's graduation time here included friends, family, misinformed counselors or advisors, and the government. Friends can encourage students to socialize to avoid loneliness or to have fun; they can also cause unnecessary drama, resulting in lost time. Families may not support a student because of insufficient finances or because they do not share the same values about the student's success in life. Family emergencies can further include the death or illness of family members. Misinformed, unmotivated or overworked counselors and advisors may lead astray students about the appropriate courses to take. Those counselors may be too busy to do the necessary research to find the proper information for the student or may lack motivation because of low salaries.[3] And then the U.S. government may impede a student's studies by not granting financial assistance because of insufficient funding resulting from a national deficit or bad public policy.

The last major cause category examined was *student*. Students may affect their own graduation time by these second-cause categories: changing majors, procrastination, health problems, being disorganized, and confusion in studies. Students frequently change college majors because of disinterest or the difficulty of the major's courses, especially in engineering, where a few do not like mathematics or are not enthusiastic because of poor-quality teachers. College students tend to procrastinate in completing their coursework because of factors such as general laziness, distraction, fatigue, lack of motivation, having better things to do, or overload. Students may have health problems because of genetics, an unsuitable environment, or a bad diet stuffed with fries and sodas. Lack of organizational skills can cause students to lose or not complete their coursework and to miss deadlines. Finally, students may be confused in their studies because they do not go to class or because they sleep in class. Unmotivated and ineffective instructors also contribute to student's confusion. Fig. 4-10 illustrates this example's CED.

According to the public-service organization FinAid (2016), college tuition increases an average of 8% every year and doubles every nine years. It is becoming increasingly difficult for students to afford college. Not all students have parents or the government funding their education. Many students must support themselves by working or joining the military. FinAid (2016) has concluded that the main root cause of students not graduating with a bachelor's degree in four years is finances.

---

[3]Apparently, this is a major concern in the higher education system.

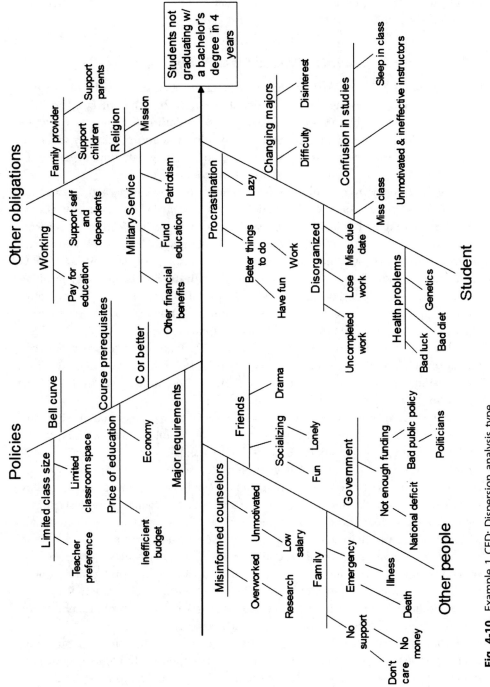

Fig. 4-10. Example 1 CED: Dispersion analysis type

The CED diagram helps in arriving to this type of conclusion by initially identifying causes. Subsequently, group meetings and idea engineering or value analysis can further analyze causes and discover the main causes.

4.5.2 Example 2: Production Process Classification Type

Group members X, Y, and Z examined the process for obtaining a bachelor's degree in engineering. According to the National Science Foundation, students who graduate with an engineering degree earn higher salaries than all other science fields, including the health field (NSF 2005). The production process classification type of CED was used to analyze this outcome. The following paragraphs spell things out in the amount of detail necessary to make an effective CED.

The first step toward obtaining a B.S. in engineering is, of course, getting admitted into college. Typically, engineering students start to think about their educational future while in high school. Most colleges require various academic accomplishments for admission: In addition to obtaining a high school diploma or General Educational Development (GED), students must also have a minimum grade point average (GPA) and a minimum (SAT) score. Students can prepare for the SAT by studying, enrolling in preparation classes, and being well-rested the day of the test. The more competitive college admissions also look for well-rounded students who participate in extracurricular activities (sports or volunteer work). Before applying to various colleges, students typically research colleges they wish to attend. Students and their parents read brochures and attend college open houses, where prospective students talk to current college students. Students then apply to their desired institutions by filling out applications, writing essays, and paying application fees.

Next, students must think about the expenses necessary to obtain an engineering degree. The most significant expense for students is college tuition, which can be paid by the student, the student's family, scholarships, and/or with government assistance. Students can pay for college tuition themselves by working or joining the military. If students decide to work while attending college, they must consider where to work and how it will affect their studies. A student's family may generously pay for college tuition or loan the student money. Scholarships are free money from private donors who wish to help students succeed; numerous scholarships are available to students but most have requirements, including financial need, achievements (academic and extracurricular activities), essays, and an application form. Moreover, the U.S. government offers financial aid to college students through loans and grants. Students apply for financial aid from the government through completing the Free Application for Federal Student Aid (FAFSA).

Other expenses besides college tuition need to be considered, including housing, transportation, food, and entertainment expenses. Students should consider whether they will live with their families or in off-campus or on-campus housing. Living with family is usually the cheapest choice for most students but may not be practical. Off-campus housing can be expensive, but other factors may lead students to decide to buy or rent a place to live. On-campus housing, such as dorms or apartments, is more affordable but not always available. Transportation is necessary for college students. Most college students want to travel by car to their destinations. However, owning private vehicles is becoming more expensive, and students have to pay for car registration, gas, maintenance, and insurance. Alternative modes of transportation include riding the bus, riding a bike or moped, walking, and—in some areas—riding rail systems. Food is a basic necessity for all, and students should research meal plans offered by colleges. Recreational expenses depend on the student's personality and should also be considered.

After gaining entry to an educational institution, students must attend classes—the most important and difficult step toward earning the B.S. in engineering. Students must be aware of general education requirements, which differ among colleges but usually include English, math, science, and history courses. Each major usually has its own specific requirements. Common engineering majors include civil, mechanical, and electrical, all of which usually require introductory classes. Elective classes can be taken to expand students' knowledge in their majors and in other areas of academia. College classes can be taken with normal grading (A through F) or as credit/no credit. Normal grading contributes to a student's GPA and total credit number. However, the credit/no credit grading option does not contribute to a student's GPA but does contribute to the total credit number.

Finally, students must meet all their college and major graduation requirements to obtain a bachelor's degree in engineering. Minimum numbers of credits in core courses, university courses, language courses, and so forth must be earned. A minimum GPA must be met. Depending on the university, student graduates may need to meet with an advisor or counselor to discuss their post-graduation plans; counselors also ensure that a student has completed all graduation requirements. All outstanding fees owed to the college must be paid off, including tuition, dormitory rent, parking tickets, overdue library fees, and graduation fees. Graduates must also fill out graduation forms. Thus, college is a long and complicated process involving early planning.

All the aforementioned causes are presented in Fig. 4-11, which illustrates the CED for this example. It is to be noted that hard work is the root factor in obtaining a B.S. in engineering, and it takes effort and dedication for students to prepare financially and academically. Earning good grades in high school is essential for getting into college; after getting in, students need to work hard and be smart to pass classes with acceptable grades. The reward after all the hard work can be for engineering graduates to look forward to obtaining a Professional Engineering license and earning the big dollars for a living.

4.5.3 Example 3: Cause Enumeration Type

Group members decided to analyze the reasons for long waiting times at restaurants with a cause enumeration type of CED. First, the group brainstormed a list of reasons why restaurants have long waiting times:

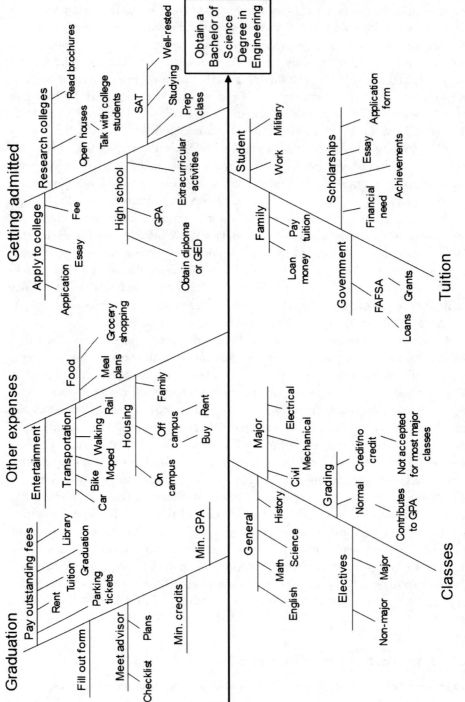

Fig. 4-11. Example 2 CED: Production process classification type

- Not enough staff
- Small kitchen
- Small oven
- Not enough stove space
- Not enough storage space
- Small refrigerator and freezer
- Not enough ingredients
- Not enough serving equipment
- Small dining area
- Poor management
- Limited restaurant hours
- No reservations
- Good reviews, very popular
- Cannot handle large parties
- Bad communication between staff
- Lack of restaurant knowledge
- Wrong orders put into kitchen
- No take-out ordering
- No prior food preparation
- Menu does not list all ingredients (allergy problems)
- Menu does not have a good description of dish
- No menu
- Holiday
- Friday night or weekend
- Laid-back atmosphere
- Good wine/alcohol selection
- Holds late reservations
- Policy requires long dining time
- Policy to not rush customers

Next, the group organized their brainstorming list into four main-cause categories for the CED: machinery, materials, methods, and people (manpower)—the four Ms:

1. Machines:
   - Small oven
   - Small stove
   - Small refrigerator
   - Small freezer
2. Materials:
   - Small kitchen
   - Not enough storage space
   - Not enough ingredients
   - Not enough serving equipment

- Small dining area
- Good wine/alcohol

3. Methods:
   - Limited restaurant hours
   - No reservations
   - Can't handle large parties
   - No take-out ordering
   - No prior food preparation
   - Menu does not list all ingredients
   - Menu does not have good description of dishes
   - No menu
   - Laid-back atmosphere
   - Holds late reservations
   - Policy requires long dining time
   - Policy to not rush customers

4. People (manpower):
   - Not enough staff
   - Poor management
   - Good reviews, very popular
   - Bad communication between staff
   - Lack of restaurant knowledge
   - Wrong orders put into kitchen
   - Holiday
   - Friday night or weekend

Then the group put together the CED, taking hours of work in the process, as illustrated in Fig. 4-12. After that, the group brainstormed even more, coming up with additional causes and third-degree causes.

The group concluded that the main cause for long waiting times at a restaurant is the owner's finances, which can make the owner cut corners. Finances control prominent causes on the CED, which were poor management and limited space. With unlimited money, an owner could hire an excellent restaurant manager and offer classes to train employees, as well as hire more employees. Unlimited money would also allow the owner to expand the restaurant to a larger space, meaning more kitchen and dining space. More kitchen space would allow for bigger appliances and more storage space for food. Larger dining areas would be able to seat more customers comfortably. However, the reality in all business is different: Finances are invariably a limited resource, and the practical realities of doing business mean that the owner has to make tough choices and difficult decisions. That said, even imagination is a limited resource, as are environmental and technological awareness. Moreover, skills, talents, and even objectives and scope of work are limited. Together, these create the business reality in every organization.

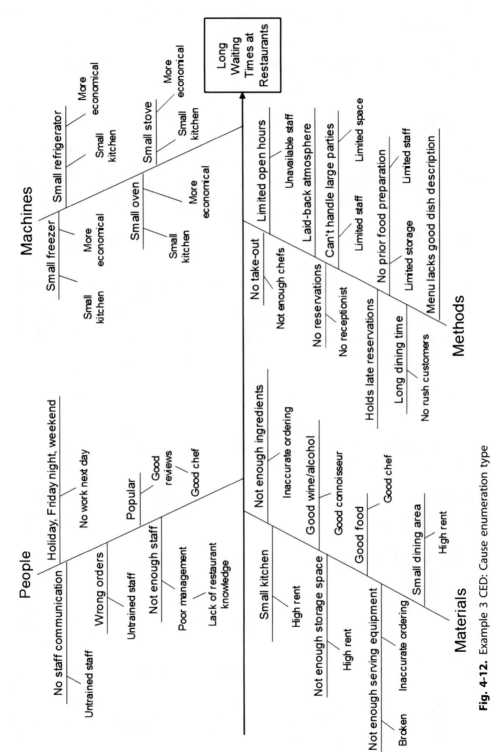

Fig. 4-12. Example 3 CED: Cause enumeration type

4.6 Exercises

The following practice problems should be completed to ensure students understand CED concepts.

Problem 1

Gather in groups of at least three people. Determine a CED for the engineering design necessary to construct a four-story engineering building. The engineering building will include classrooms, lecture halls, laboratories, and offices. The building will be built on an empty gravel lot next to a stream. Use the dispersion analysis or cause enumeration type of CED.

Problem 2

Convert the CED obtained in the previous exercise into a production process classification type.

Which diagram is easier to understand? Is there a noticeable difference between the diagrams? Is the root cause of the diagrams the same? Which diagram is easier to work with?

References

American Society for Quality. (2010). "Kaoru Ishikawa." <http://www.asq.org/about-asq/who-we-are/bio_ishikawa.html> (Jan. 17, 2017).

Brassard, M., and Ritter, D. (1994). *The memory jogger II: A pocket guide of tools for continuous improvement and effective planning*, 1st Ed., GOAL/QPC, Salem, NH.

FinAid. (2016). "Tuition inflation." <http://www.finaid.org/savings/tuition-inflation.phtml> (Jan. 20, 2017).

Goodden, R. L. (2009). *Lawsuit!: Reducing the risk of product liability for manufacturers*, Wiley, Hoboken, NJ.

Ishikawa, K. (1976). *Guide to quality control*, Asian Productivity Organization, Tokyo.

NSF (National Science Foundation), Division of Science Resources Statistics. (2005). "Recent engineering and computer science graduates continue to earn the highest salaries." <http://www.albany.edu/faculty/tam/fall/Employment%20of%20CS%20&%20IS%20grads%20(nsf06303).pdf> (Jan. 19, 2017).

Omachonu, V. K., and Ross, J. E. (2004). *Principles of total quality*, 3rd Ed., CRC Press, Boca Raton, FL.

Simon, K. (2010). "The cause and effect (a.k.a. fishbone) diagram." <http://www.isixsigma.com/library/content/t000827.asp> (Jan. 24, 2010).

U.S. Department of Education, National Center for Educational Statistics. (2009). "The condition of education 2009 (NCES 2009-081), Indicator 22." <https://nces.ed.gov/pubs2009/2009081.pdf> (Jan. 20, 2017).

CHAPTER 5

Function Analysis System Technique—Structuring Uncertainty

5.1 Introduction

Value management encompasses the gamut of proven methods for the analysis and valuation of processes, products, and projects. It aims to optimize processes to reduce expenditure and maximize worth. As a profession, value management has evolved with each newly proved theory, whether it is an addition to an already established and similar method, a groundbreaking new method, or the integration of two or more proven methods. Its basic concepts stem from value analysis and value engineering; value management has become an umbrella discipline that defines not only the processes but also the terms for the profession (Kaufman 2008). It has directly influenced the business world—and, more profoundly, the manufacturing market—by optimizing methods for cost reduction while retaining essential product functionality.

This chapter will define and expand on one of these methods, the *Functional Analysis System Technique (FAST)* method, and will explain how value management can be used to structure the operation of tasks and reduce uncertainty. FAST will first be analyzed by defining and using concepts already established within value methodology. This is necessary because FAST was developed from value methodology, although it has become a more specific technique that includes a wide range of practices. This chapter will also chart the arc of FAST's development and applications (Wixson 1999), demonstrating basic principles of value management.

This chapter is included because FAST diagramming helps both to structure uncertainty in product classification and to enable engineers to arrive at decisions regarding the aims set out in product improvement. In addition, value engineering with FAST diagramming is a step up in engineering science, a means to make wonderful improvements by placing innovation on a structured path to success.

Examples are drawn from people's daily lives, mainly because this makes it easier to understand than would the use of high-flying engineered examples and constructs that may not resonate with every reader.

5.2 Background of the Functional Analysis System Technique (FAST)

As we define and outline the development of FAST, a deeper understanding of its history will help the reader better understand how it contributes to the advantageous features of value management and how it is unique in the family of professional methods that help structure uncertainty. FAST is one of the functions within the *function analysis* phase, which is a step in the job-plan technique of value methodology, which is in turn embodied in value management; these relationships are conceptualized in Fig. 5-1.

Value engineering was not an established system until 1947, when Lawrence D. Miles, an engineer at General Electric., started researching substitute materials and processes that could minimize the cost of manufacturing without sacrificing the main function of the products (Lawrence D. Miles Foundation n.d.). Miles formalized his methods in several publications. His first book, *Techniques of Value Analysis and Engineering*, was originally published in 1961, with its third and last edition published in 1989. Miles said that value methodology is "a disciplined action system, attuned to one specific need: accomplishing the functions that the customer needs and wants, whether these functions are accomplished by hardware, service a group of people, professional skills, administrative procedures, or other at the lowest cost" (Miles 1989, p. xvi).

The key to this statement is understanding the central, most fundamental concept in value methodology: that value is defined in terms of function, in other words, the ability of the product to perform the tasks for which it was designed. Miles (1989, p. 4-5) first defined *value* by stating "A product or service is generally considered to have good value if that product or service has appropriate

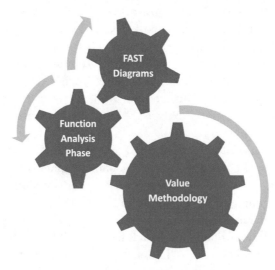

Fig. 5-1. Hierarchy for FAST diagrams in value management

performance and cost. It can almost truthfully be said that, by this definition, value can be increased by either increasing the performance or decreasing the cost." In modern times, value is defined by the following commonly known relationship, where performance of the service or product is understood as the functions it performs:

$$Value = \frac{Function}{Cost}$$

Function represents the basic purpose of each expenditure, whether it be for hardware, the work of a group of people, a procedure, or anything else needed to accomplish function. It is necessary to establish the language of function and then work within this language. *Cost* represents the expenditure needed to accomplish the function and all related factors.

Value methodology can be broken down into the essential *job plan* or *management* phases defined by the value methodology organization, SAVE International, as shown in Fig. 5-2 (SAVE International 2007, Miles 1966). "The value methodology is a structured, disciplined procedure aimed at improving value. That procedure is called the Job Plan. The Job Plan outlines sequential phases to be followed which support team synergy within a structured process, as opposed to a collection of individual opinions." (SAVE International 2007, p. 12).

The steps given in the job plan must be strictly followed, as they are central to the success of the value engineering method. Following this disciplined sequence of planning events vastly increases the chance of success at product improvement or meeting other stated objectives. By formalizing the structure of function analysis, cost reduction, or product improvement, it is much easier to cover all engineering and related factors that make up the system that we are addressing. Consequently, SAVE International guidelines for value engineering exercises mandate that the value management job plan be adopted (SAVE International 2007).

5.3 Function Analysis

The function analysis phase is unique in its purpose and techniques, standing apart from traditional methods and approaches in the value methodology phase structure. It is a cornerstone of value methodology because it defines the sole purpose of the job-plan exercise (Kaufman 2008). Function analysis itself consists of three main phases—including the FAST method—depicted in Fig. 5-3 (Stewart 2010), which also lists the steps that must be accomplished during job planning in the value-management exercise.

Although the steps and phases of the job plan may appear simple to the engineer and layperson alike, it is emphasized over and over in value-engineering literature that adherence to this disciplined scheme can make the difference between success and failure. Being dedicated and committed to this method is necessary to ensure that the exercise stays on track.

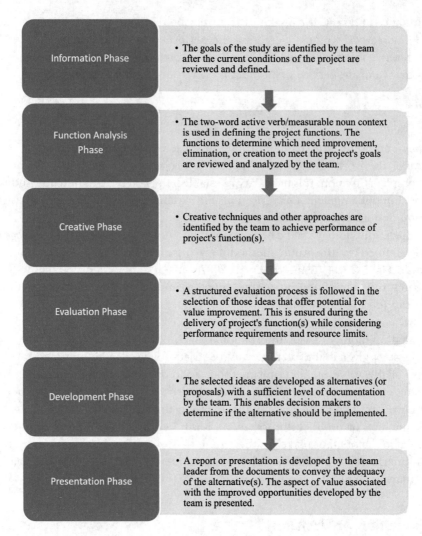

Fig. 5-2. Job plan phase tier for value methodology

5.3.1 Defining Function

Function can be defined, in simple terms, as what makes a product work. When Miles (1989) was first presented with the problem of cost reduction in the 1940s, he quickly identified the core of the problem as function. It is the reason that a consumer buys a product and how well the product sells. Function comes first, cost next. In his own words, in 1989 he stated:

"The heart of the situation is "the customer wants a *function*." He wants something done. He wants someone, perhaps himself, pleased. He wants something enclosed, held, moved, separated, cleaned, heated, cooled, or whatever under certain conditions and within certain limits; and/or he

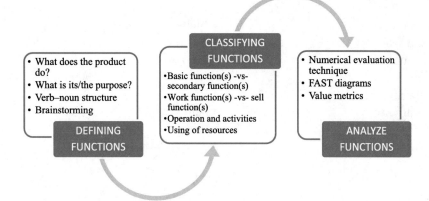

Fig. 5-3. Function analysis phases

wants a shape, a color, an aroma, a texture, a sound, a "precious" (costly) material, or whatever to bring pleasure to himself or others he wishes to please. That is *all* he wants. That is *all* he cares about.' Miles 1989, p. 25)

As the first step in function analysis, defining the function of a product or a project is critical. In fact, Kasi (1994, p. 14) notes that "Defining functions and understanding them differentiates value engineering from other cost-reduction techniques" and so needs to be focused upon to develop the value engineering job plan. In value methodology, defining and determining function can only be accomplished through careful consideration of the product needed or of how consumers will use the product for their needs (Stewart 2010). Hence, the first, simplified principle in defining function is the use of the *verb-noun* discipline to make certain that all concerned really understand, in a succinct way, the fundamental nature of the function. That is to say, after a simple rewording to frame the paradigms more clearly, the task becomes quantifiable in terms of definition and cost.

The verb should be in active voice and answer the question, *What does it do?* Answering this question will put the focus on the function of the product and allow a clearer approach to its functional use and analysis. In determining this part of the function definition, cost reduction is also introduced at this point (Miles 1989). It helps the project group or designer determine what the product does and how it can be accomplished at a lower cost. The active verb is immediately followed by the question, *What is this verb applied to?* This is answered by the *measurable* noun that will describe exactly what is acted on (Stewart 2010). The noun must be in measurable terms, as this will help any specific calculations during the evaluation of this action, such as determining the cost of the full exercise. The combination of these two words for function definition will provide a work function, which is called the *verb–noun abridgement* (Miles 1989). The following lists include examples of active verbs and measurable nouns.

Active verbs:

- Apply
- Approve
- Change
- Close
- Open
- Condense
- Conduct
- Elongate
- Emit
- Fasten
- Guard
- Handle

- Incite
- Include
- Interrupt
- Resolve
- Revolve
- Safe
- Scale
- Situate
- Sort
- Thwart
- Transfer
- Treat

Measurable nouns:

- Air
- Current
- Data
- Density
- Energy
- Fluid
- Force
- Particles
- Protection
- Solids
- Sound
- Speech

- Friction
- Gear
- Light
- Liquid
- Noise
- Object
- Paint
- Switch
- Torque
- Vibration
- Voltage
- Volume

Although defining a function in terms of an active word and measurable noun might appear trivial, because we all learned about verbs and nouns in fourth or fifth grade, it is not always simple for many processes and can consume hours of mind-wrenching word combinations and many revisions until the group is satisfied with the final selection. Discouraging as this may be, there are instances when it is more practical to use a less forceful verb or a noun that is not easily measured or cannot be quantified. The following lists include examples of useful less forceful verbs and unquantifiable nouns.

Passive verbs:

- Permit
- Correspond
- Suggest
- Reveal

- Increase
- Preserve
- Handle
- Propose

- Exhibit
- Improve
- Assist

- Bestow
- Offer
- Represent

Unquantifiable nouns:

- Aesthetics
- Sanction
- Splendor
- Ease
- Expediency
- Posture
- Shape

- Exterior
- Conserve
- Delight
- Joy
- Agreement
- Position
- Balance

It is not uncommon to have an abridgement combination consisting of an active verb and a measurable noun or a less forceful verb and an unquantifiable noun. However, the focus should be on having a working function that is described using the active verb and measurable noun abridgement; in this way, the function is described in measurable terms so that its highest definition can be achieved. Also, if further analysis is conducted, cost can be defined, because quantities are easily provided and measured. For example, consider a power line to a residential building. One could describe this function using the verb–noun abridgement *provide service.* Although this is a true description of the overall purpose of the power line, it is difficult to measure the *service* provided.

However, one can reword this function in quantifiable terms, such as using the verb–noun abridgement *conduct current.* This verb–noun abridgement better describes what the power transmission line function to the home is; if a cost should be applied, it is easily measurable from the quantity of amperes used. Other related costs include the applied cost of actually performing this action, as expressed in the active verb *transmit* and may be measured in terms of materials, equipment, and labor costs to operate or undertake this action.

Redefining a function with a verb–noun abridgement will allow designers or a multidisciplinary project team to "build a shared understanding of the functional requirements of the project" and, as a result, allow them to "identify where opportunities for value improvement exist in the project" (SAVE International 2007). Stewart (2010, p. 149) expressed the advantages of using working function abridgements to define functions:

- It avoids combining functions and defining more than one simple function. By using only two words, you are forced to break the problem into its simple elements.
- It forces conciseness by defining a function in two words. If with two words, the functional component is still too large to be defined or to be understood by the team, then this is a sign to drill deeper and to produce a more detailed two-word definition.

- It aids in achieving the broadest level of dissociation from specifics. When only two words are used, the possibility of faulty communication and misunderstandings is reduced to a minimum.

The value and power of using a verb–noun combination cannot be underestimated. It is a powerful mechanism that forces the designers to consider deeply what the functions are and to focus their minds on those functions. Moreover, it helps narrow down the performance requirements of the product, which reduces the risk of improper problem identification. This will eventually result in the proper design being delivered and consequently improve marketability.

5.3.2 Classifying Functions

In the field of value engineering, functions are classified as follows:

- Basic functions
- Secondary functions
- Higher-order and lower-order functions
- Assumed functions
- All-time functions, and
- Dependent and independent functions

5.3.2.1 Basic Functions

A basic function is the necessary purpose of a product or project. These functions reveal the usefulness of a product or project and the reason for its use or existence (SAVE International 2007). Basic functions should be able to answer the question, "What must it do?" (Stewart 2010, p. 150). For example, a basic function of a wall clock is to *indicate time*. All other functions are secondary or supportive of this basic function; for example, a decorative or intricate clock face would have, in addition to its basic function, the function of providing aesthetic pleasure for the buyer, in other words, to *provide aesthetics*. The basic function is located just to the right of the left-side scope line of the FAST diagram, before the higher-order function. The general rules and grammar for arriving to basic functions are found in Fig. 5-4.

5.3.2.2 Secondary Functions

Secondary functions support the basic functions. Their main contribution is to enhance the sole purpose or basic function of the product or project, and they "may or may not be essential to the performance of the basic function" (Stewart 2010, p. 150). It is also important to identify secondary functions to quantify related costs, which usually make up the majority of the product or project cost. During further cost analysis, designers and project teams can reduce costs of a project's secondary functions so that the basic function value is maintained more inexpensively.

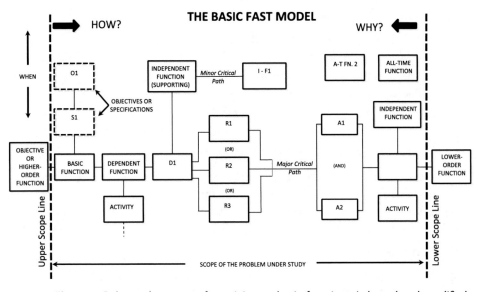

Fig. 5-4. Rules and grammar for arriving to basic functions (adapted and modified from Stewart 2010 and Kaufman 2008)

5.3.2.3 Higher-Order and Lower-Order Functions

Higher-order functions are defined by the needs that basic functions are expected or required to meet or satisfy. These functions answer the question, "Why is it needed?" for a product or project. They also help define the purpose of a product and relate it to a consumer demand or need. For instance, the higher-order function of installing a water tap may be to *drink water*, which may have the further higher-order function to *quench thirst* or *maintain health*. Higher-order functions are located at the far left of the FAST diagram and follow the basic function.

Lower-order functions, located at the far right of the FAST diagram, indicate the starting point of the scope of work being addressed. Hence, a possible lower-order function of installing a water tap might be to *feel thirsty* or *feel tired*.

5.3.2.4 Assumed Functions

Assumed functions are related to either the basic or secondary functions but are not included in the analysis because they are outside the scope of work. For example, when considering a smartphone, one could say that its basic function is to *transmit data*. Data can be measured as the number of bytes sent via text, as file size, or as the number of phone calls made. *Transmit* is an active verb that answers the question, "What does it do?" An assumed function of a smartphone would be considering the amount of coverage available to carry out its basic function. The available coverage within a usage area is not directly related to the basic function, but it is an assumed function because it is needed to carry out the basic function of a smartphone.

5.3.2.5 All-Time Functions

At the upper right of the FAST diagram, there is a position for an all-time function. The purpose of this function is to understand additional features of the process, because when analyzing processes, it is useful to know as much about them as possible. For example, the all-time function of a highway could be defined as *transmit traffic*; for an office, it could be *undertake work.*

5.3.2.6 Dependent and Independent Functions

These two types of functions represent whether there are dependencies going into or coming out of a function (dependent function) or whether a function can stand alone as an end in itself outside of the critical path (independent function).

5.3.2.7 Critical Path

This is the main path of the process, which shows progression from earlier functions to later functions, in sequence, spanning the scope lines and going from lower-order functions to the basic function to higher-order functions. This critical path must not be confused with the altogether different one used in the critical path method (CPM). In the FAST model, the critical path is the most important path on which the value engineers must focus, rather than on peripheral functions such as the all-time function.

5.3.2.8 Analyzing Functions

As shown in Fig. 5-3, the third and last step in the function analysis method is analyzing functions. This consists of *levels of abstraction,* the *numerical evaluation technique,* the FAST method, and *value metrics* (a derived form of the FAST method). The other techniques will not be discussed in this book, but it should be noted that their contributions are as important to this part of function analysis as that of the FAST method. However, we are concentrating on FAST, because it has become the most common method of function analysis.

Determining basic functions is a difficult process. Variation in defining a basic function can be high enough to dictate the ease or difficulty of the entire function. For instance, most people would say that the function of a key is to *open lock.* Although this is correct, in a sense, and provides a good starting point, the question should then be asked how a key opens a lock. When we break down the actual motions or actions taking place, the key itself cannot open a lock unless something *transmits torque* (i.e., provides power to turn the key in the lock), thus allowing the initial assumption to be true. As stated previously, it is important to define a function in measurable terms. In this case, the noun *torque* is measurable, whereas the noun *lock* is not.

Again, functions are simple descriptions of each activity and consist of two-word verb–noun descriptions. These are intended to minimize confusion with organizational functions such as engineering and finance. Bartolomei and Miller (2001) provided another example of verb–noun descriptions for an automobile control system. The basic function of an automobile is to transport people from one place to another. We must not dwell only on how the automobile is used for transportation; other functions, such as how the environment will perform, are also important. Whereas the car is being driven, the verb–noun descriptions—not combinations—can be as follows:

Component	Verb	Noun
Monitor environment	check	witness
Instruments	watch	object
Control direction	organize	rule
Control speed	run	power
Passenger comfort	relieve	luxury
Control visibility	clarity	profile

Here, driving the car is a higher-level function. The other functions determine the means to achieve the higher-level function.

5.4 Function Analysis System Technique and Value Methodology

The most commonly used technique for function analysis is the FAST method. (The U.S. Navy prefers to use the name Functional Analysis Concept Design [FACD], although it is completely identical to FAST [Stocks and Singh 1999].) Expanding on Miles's value methodology concept, FAST was developed by Charles W. Bytheway in 1964 and first introduced at the 1965 National Conference of the Society of American Value Engineers (Johnson 2013). The introduction of FAST to value methodology was a tremendous contribution to function analysis. FAST may be used in any type of setting or for any situation as long as the functionality is defined. It provides essential techniques for function analysis and helps engineers to understand risks by understanding prices, functions, and processes, but it also comes with limitations that can prevent progress without proper understanding of its use and purpose. FAST can also provide an advantage for function analysis by bringing together a team—regardless of their knowledge, education, and technical background—which can propose and promote a fuller understanding of the problem or project at hand (SAVE International 1998). It is with this purpose that FAST becomes an excellent tool for handling risk and uncertainty, especially in times of crisis or disaster.

5.4.1 The How–Why Concept

Bytheway (2005) described developing the FAST method as applying the known function analysis method in a manner that uses the *how-why* concept. In other words, most of what was stated previously about the function analysis phase of value methodology is now combined and arranged in a logical sequence by asking the questions *how* and *why* and by arranging them using the FAST diagram. The FAST diagram is arranged so that the question *how* is read from left to right and the question *why* is read from right to left. To quote Bytheway, "When we ask 'How' we are looking for solutions and moving to lower levels of opportunity. When we ask 'Why' we are looking for reasons and moving to higher levels of opportunity" (U.S. Department of the Interior 2008, p. 22). The FAST model is described in Fig. 5-4 and Table 5-1:

Table 5-1. Features of a Basic FAST Model

A.	Scope of the problem under study	The area where the product or project is bound for study by the designer or project team.
B.	Output/objective or higher-order function(s)	These functions are of the highest order and placed on the far left in the FAST Diagram. The functions that have higher standings are placed to the left of the scope line.
C.	Input/lowest-order function	The lowest-order function is placed on the far right of the FAST Diagram, to the right of the scope line. It helps the system to achieve the goal.
D.	Basic function	The basic function is found immediately to the right of the highest-order function and describes the purpose of the product/project.
E.	Concept	These secondary functions are immediately to the right of the basic function and represent the supporting role or functions required to achieve the basic function and, in turn, the highest-order function.
F.	Object or specifications	These are parameters required to achieve the highest-order function. Parameters may not be functions by their nature.
G.	Critical path function(s)	These functions are found along the *how-why* concept line.
H.	Dependent functions	Dependent functions are found immediately to the right of the basic function and all that follows. These are secondary functions and dependent on the function to their immediate left for their functionality.
I.	Independent (or supporting) functions	These functions are isolated and do not depend on another function to perform their role. They are located directly above the critical path function(s).
J.	Function	Function is expressed in verb-noun form, and it is an end or a purpose for which an activity is performed.
K.	Activity	This is the method selected to perform a function or group of functions.

The uniqueness of FAST is the ability to break down the process of identifying the basic function using a logical sequence. Using the how-why concept is illustrated in the general FAST diagrams for a few cases in manufacturing shown in Fig. 5-5 (Crow 2002).

5.4.2 When Logic

The *when* direction in the basic FAST diagram indicates the logical flow of the cause and effect of a function, not the time reference. The *when* function is drawn vertically, in either the up or down direction. Using *when* may also create an

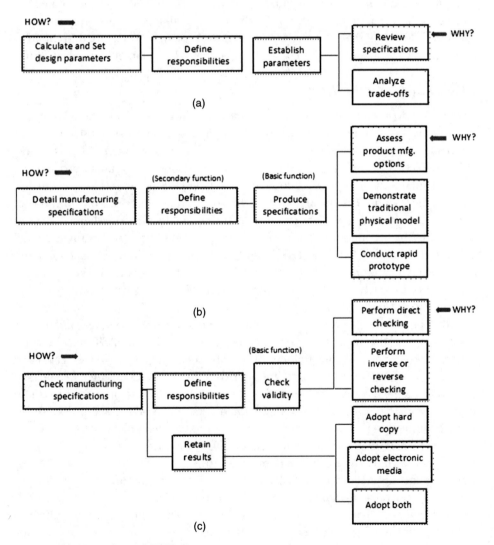

Fig. 5-5. General FAST diagrams

alternative and/or minor path that becomes a subsystem in the FAST diagram. This enhances the performance of some functions and may better explain the overall function of the project and/or product (Stewart 2010). Unlike functions in the critical path of the diagram, changing the functions in a subsystem will not necessarily affect the basic function(s).

5.4.3 And/Or Logic

The *and/or* logic in a FAST diagram indicates the way in which functions are connected. For example, *and* indicates a division or a fork between functions that may have equal weight and effect in answering the *how-why* questions. This is indicated by a split between functions, both paths of which are to be followed. The *or* connection indicates a choice in function path or an alternate function path whose importance is equal to or less than that of the *and* path. And/or logic may also be used in the *when* direction. Fig. 5-4 illustrates the *and* and *or* concepts (Kaufman 2009, Stewart 2010).

5.4.4 FAST Procedure Tree

Charles Bytheway wrote several articles outlining his thinking process about FAST and its development. In one of these articles, Bytheway (2005) emphasized that although he created the technique, he did not finish FAST diagrams for or apply them to many of his projects. The reason is that there are many ways one can go about producing a FAST diagram, including using adopted and established methods. However, in this chapter the emphasis is on highlighting the originator's steps for the purpose of understanding the method's origin. Other methods are detailed elsewhere, such as in publications by the American Society for Testing and Materials (ASTM International 2005), but the one by Bytheway (2007) is more detailed and extensive. The following are Bytheway's recommended steps in producing a FAST diagram, also known as the FAST procedure tree, as detailed extensively in his book (Bytheway 2007). This chapter will not go into further details beyond the outline shown in Table 5-2.

Although the FAST procedure tree may seem tedious and elaborate—and it often is—it is by far the most thorough method of developing a FAST diagram and structuring the risks and uncertainties of the problem. Of course, project teams and designers may modify the tree to suit their needs without compromising the main goal of the activity. Each of the steps detailed by Bytheway for the FAST procedure tree is an expanded or supporting part of the function analysis process outlined previously. In fact, the thirteen steps make up a more elaborate and detailed version of the three-step process of function analysis, which includes all of the critical steps: *define function*, *classify function*, and *evaluate function* (Stewart 2010). The FAST procedure tree not only replicates the advantages of the FAST method within function analysis but also can contribute more by providing a methodical process for each step. It is a readily available blueprint that may be altered or modified within its own steps and limitations without compromising the overall goal of the project, or expanded to cater to each unique situation, product, or project.

Table 5-2. Procedures to be followed in FAST modeling:
The FAST procedure tree

	Procedure
1.	Select project
2.	Select participants
3.	Identify initial functions
	Supply names/drawings (as applicable)
	Share initial functions
	Determine diagram type
4.	Determine initial basic function
5.	Develop higher-level functions
	List higher-level functions
6.	Determine basic function
7.	Develop primary-path functions
	• Make function list
	• Post basic function
	• Apply *how* logic
	• Apply *how-why* logic
	• Verify tree logic
8.	Evaluate remaining functions
	• Develop function clusters
	• Identify supporting functions
	• Merge primary-path clusters
9.	Identify/add supporting functions
	• Use *when-if* logic
	• Merge supporting functions
10.	Develop secondary-path functions
	• Apply *how* logic
	• Apply *how-why* logic
	• Merge supporting function clusters
	• Investigate remaining functions
	1. Apply *why* logic
	2. Merge/cross out clusters
11.	Brainstorm higher-level functions
12.	Generalize functions
	Brainstorm generalized functions
13.	Develop and brainstorm undisclosed functions

Source: Bytheway (2007).

Example 1: Purse Zipper

The basic function of a purse zipper is to fasten the purse flaps and secure the contents of the purse; its higher-order function is to *secure* the *purse*. Now, functions are identified and arranged in an order that will allow for a sound analysis using the

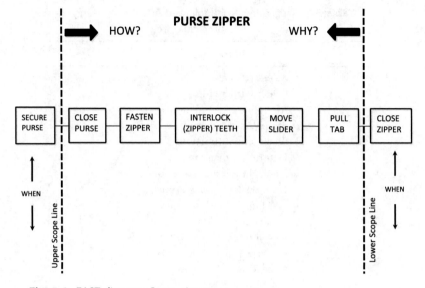

Fig. 5-6. FAST diagram: Purse zipper

how-why concept. The how-why concept can be understood after goals are identified for the problem. Fig. 5-6 illustrates the FAST diagram of the purse zipper, as applied to the processes of value engineering.

Moving from right to left on the diagram answers the question of *why* a function is performed; moving from left to right answers the question of *how* the function is performed. So the answer to "Why do we *pull tab*?" is "Because we want to move the slider"—we cannot move the slider unless we pull the tab. Similarly, the answer to "How do we *move slider*?" is "By pulling the tab"—pulling the tab is necessary to move the slider.

We start with the lower-order intention of closing the zipper and end with the higher-order goal of securing the purse, with the basic function of the zipper being to *close* the *purse*. The sequence has a cause-and-effect relationship; changing the placement of a function will affect the functions to its right and to its left. A reversal of the how-why concept will allow the testing of the first word of the logic, answering the *how* question by reversing the direction we used to answer the *why* question. For example, if the function *pull tab* is removed, *move slider* will not be functional in either the how or why questions.

The *when* direction in the FAST diagram indicates the processes involved in the cause and effect of a function. Using *when* may also create an alternative or minor path as a subsystem in the FAST diagram.

Unlike functions in the critical path of the diagram, the functions in a subsystem can be changed without necessarily affecting the basic function. The procedure for handling the interacting functions has the following steps:

1. Arrange the functions in a suitable order for the purpose of evaluation.
2. Evaluate the function at the top of the list as though it were a single function.
3. Evaluate the second function in the arrangement under the conditions that are assumed to have provided solutions for the first function.
4. Use the same approach for the remaining functions. Upon completion, it is evident the functions are compatible with each other.

In this example, the functions along the critical path include

1. Close purse (lower-order function),
2. Pull tab,
3. Move slider,
4. Interlock zipper,
5. Fasten zipper,
6. Close purse (basic function), and
7. Secure purse (higher-order function).

As we look at the work of the higher-order function *secure purse*, we see that the work would be futile if the other functions were not performed in the process of closing the purse.

Example 2: Paper Scissors

Let's consider the functions involved in a paper scissors model. The main function of paper scissors is to change the shape and size of paper. In this example, the higher-order function is to work with scissors to *reshape paper*. The first step is to identify the functions. As we identify the functions and arrange them in order, we shall look at the analysis using the how-why concept, which can be understood after we identify goals for the problem.

Moving from right to left, the process starts with a need (changing the shape and/or size of the paper) and ends with the function that will achieve the goal, *reshape paper*. The line of thinking can be read either from left to right or from right to left. Reversing the how-why concept will allow the testing of the first word of the logic, answering the how question by reversing the direction in which we answered the why question. For instance, why do we *insert fingers*? It is because we must hold the handles. Conversely, how do we *hold* the *handles*? By inserting our fingers.

The FAST diagram is illustrated in Fig. 5-7. From left to right, in detail: How do we cut paper? By using scissors. How do we use scissors? By holding the handles. How do we hold the handles? By inserting fingers. How do we insert fingers? By using the finger holes. How do we use finger holes? By squeezing our fingers. How do we squeeze our fingers? By moving the handles and crossing the blades of the scissor and activating the pivot screw. How do we cross the blades and activate the pivot screw? By working the scissors.

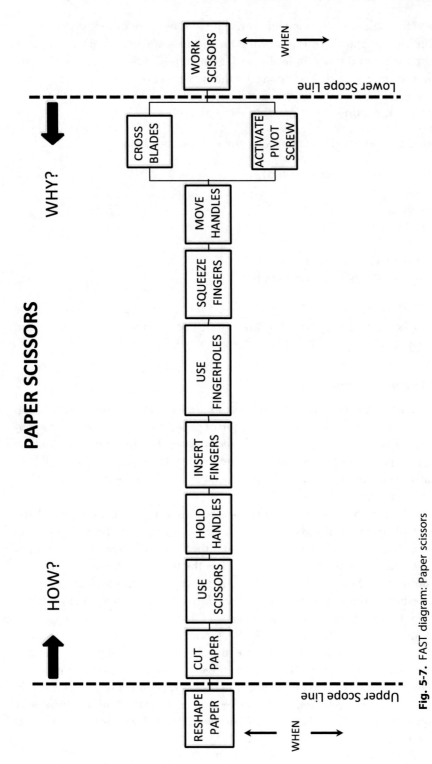

Fig. 5-7. FAST diagram: Paper scissors

In further detail, the logical flow helps us to understand the functions in the sequence:

1. Reshaping paper (the *reshape paper* function) is the final objective in this model. How do we accomplish that? We cut paper with scissors.
2. To reshape paper we must cut it (*cut paper* function). How? We use an available set of scissors.
3. How precisely do we *use scissors*? We insert fingers and enable the *hold handle* function.
4. How do we hold the handle? We employ the function *insert fingers*.
5. How do we insert fingers? We enable the function *use finger holes*.
6. How do we use the finger holes? We *squeeze fingers* to move the scissors blades.
7. How exactly do we squeeze our fingers? Not in just any direction, but in a way that enables the *move handles* function.
8. How do the handles move? There are two components. The scissors' *cross blades* are attached via a pivot screw that moves the blades in either the up-and-down or the forward-and-backward direction. This enables the blades to cut paper.

In the end, the task at hand—the *work scissors* function—helps us to understand the need for a pair of scissors to be made to work to achieve the objective of reshaping paper.

The reverse process is also true as we go from right to left on the diagram, this time answering the *why* question. Essentially, the how is the inverse of the why. The purpose for which we are undertaking a function is answered by asking why we are doing it, moving from right to left in the FAST diagram. The functions are immensely dependent in this scenario, as we find that reshaping paper is done by working the scissors on paper. Thus, the sequence has a causal relationship, so that changing the placement of a function affects the functions to its right and left. For example, if the function *use scissors* is removed, then there is no use for the next function, *cut paper*. Hence it will not be functional in either the how or why question.

Example 3: Coconut Water

Let us consider one more example: coconut water. The main higher-order function of the *coconut water* model is taken to *satisfy thirst* for which the basic function is *drink coconut water*. As the functions are identified and arranged in order, we shall look at the analysis using the how-why concept.

After many hours of brainstorming—discussions; taking in ideas and shredding some; going iteratively back and forth on suggestions and reasons; discussing pros and cons, functions, and purpose; adding and erasing functions on the critical path on a whiteboard numerous times; bringing back some erased functions and weaving the functions together; and seeking to optimize the length of the FAST diagram to keep it succinct, yet complete—the functions the team settled on were *retrieve coconut*, *strike coconut*, and *create opening*. The FAST diagram was developed as illustrated in Fig. 5-8.

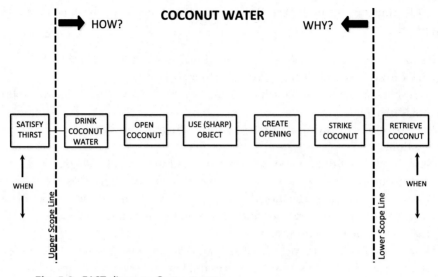

Fig. 5-8. FAST diagram: Coconut water

Moving from right to left, the process starts with a need and ends with the function that will achieve the goal *satisfy thirst*. The lower-order function is to retrieve the coconut in order to fulfill the higher-order function of satisfying thirst; the basic function is essentially to drink the coconut water. In this case, without drinking the coconut water the thirst will not be satisfied; conversely, to satisfy the thirst, one must drink the coconut water. Hence, we see that the functions are interconnected. For example, if the function *drink coconut water* is removed, then the adjacent function *open coconut* is not relevant. Similarly, as we look from right to left, the function *strike coconut* has no use without the function *create opening*; hence, without each other, neither function can be useful in either the how or the why direction. Adjacent functions supplement, reinforce, give meaning to, and are necessary for their immediate neighbor functions. Hence, the entire critical path is a string of necessary functions that must exist for the process to be explained and accomplished at all.

As we examine the model from left to right—the *how* direction—the logical flow helps us to understand the functions in the sequence. The underlying functions between these two major functions follow a certain sequence that demonstrates the how and why logic processes, and the intermediate functions are dependent on each other as in the and/or logic process.

1. The thirst satisfaction (*satisfy thirst*) function is the final objective in this model. This is achieved by *drink coconut water*.
2. The *drink coconut water* function is used to fulfill the goal of quenching thirst.
3. To drink coconut water, the coconut needs to be opened, so the function here is *open coconut*.

4. To open a coconut, one needs to use a sharp object, so the function here is *use sharp object.*
5. To use a sharp object, one needs an opening where to insert the sharp object, and so one needs to create an opening, so the function here is *create opening.*
6. One can only create an opening by striking the coconut, and
7. One strikes the coconut only after retrieving it.

5.5 Conclusion

The FAST method is an essential tool for value management. It structures a method of identifying the functions and purposes of a product or project and arranging them in meaningful relationships. Function is the most important factor in value determination using function analysis. Without determining the function of a product or project, its value may not be optimized to its maximum level, and opportunities for cost reduction may be lost. So far, the FAST method has been able to stand the test of time because of its simplicity and ability to fine-tune the most essential part of value determination through function analysis. Thus, the useful and advantageous properties of the FAST method can be applied as an essential tool for a wide spectrum of professions and projects, including risk analysis. Value analysis and FAST techniques help in problem solving and decision making by analyzing thought processes in detail.

5.6 Exercises

Problem 1

Energy is an important and basic infrastructure required for the economic development of a country. Globally, many countries are turning to renewable energy sources to fulfill their energy needs. Renewable energy sources are gaining importance because of their ability to minimize climate change effects such as global warming. Within the next 30 years, renewable energy sources may make up 50% or more of total energy use and make the world greener. Energy security is imperative for the sustained growth of an economy, and renewable energy sources are seen as the most ideal alternative for filling energy demand without causing damage to the world's environment. Use value engineering methods and techniques to solve each of the following independent issues.

1. What is the most feasible way to use renewable energy resources?
2. Select the right energy source for a specific application, such as electric-power generation.
3. Design an energy system for a given specification based on current, commercially available technologies.

4. Explain the factors to be considered when selecting a site for a wind-power plant.
5. Outline the overall benefits and analyze the cost-benefit ratio.

Problem 2

An engineering company manufactures radio frequency (RF) generators with mass spectrometers for applications in science and the petrochemical and medical industries. The generator supplies 2 kW of RF power at 27 MHz to produce argon plasma at temperatures of 10,000 K to allow grating separation for a sensitive camera to analyze. Design and production is expected to start in 2018. Production will be 25 units per week, with regular cost-reduction initiatives. The company is planning to have a new replacement design available in 2022. Use value engineering methods and techniques to solve each of the independent issues that follow.

1. What changes to the design are required to protect the devices from excessive voltage?
2. What improvements are needed in machine setup and testing?
3. What improvements are required in metalwork, and what is the cost involved in design?
4. Outline the overall benefits and analyze the cost-benefit ratio.

Problem 3

A technology solutions group uses state-of-the-art infrastructure for heavy-engineering repair services. They have a top-of-the-line repair facility that they want to close, raising capital by liquidating the assets. Most of its activities were transferred into existing businesses, but they could not find a solution quickly enough to relocate the repair facility. The facility was spread over an area of 8,000 sq. ft of workshop and component/spares stores, and the group wanted to reduce that to 4,500 sq. ft. However, the company faced major challenges related to the following:

- Physically transferring all the equipment,
- Creating a seamless transition over six months so that customers would not experience performance downgrades,
- Redesigning the space for this purpose, and/or
- Reducing space requirements.

The company seeks solutions to the following independent issues; use value engineering methods and techniques to solve each one:

1. How to transfer all the repair equipment
2. How to engage staff already working in the unit

3. Logistics and personnel plans
4. Benefits of reducing space and assets, and what are the indirect savings
5. Complete this relocation exercise in six months' time.

Problem 4

An electrical utility industry manufacturer wants to reduce the manufacturing cost of a product by 20% to have an advantage over its competitors. The company wants to formulate an action plan in order to meet this objective. The current issues were analyzed thoroughly by a team consisting of management and engineering staff members. The team defined the actual requirements and came up with an action plan for realizing potential cost savings. Use value engineering methods and techniques to solve each of these independent issues:

1. What are the current market needs and the limitations of the product design?
2. What is the manufacturing cost?
3. What are the selling features and functions?
4. How do the adaptability and cost of this product compare to other models?
5. How easy it is to install and maintain?

References

ASTM International. (2005). "Standard practice for constructing FAST diagrams and performing function analysis during value analysis study (ASTM E2013-12)." <http://www.astm.org/Standards/E2013.htm> (Jan. 27, 2017).

Bartolomei, J. E., and Miller, T. (2001). "Functional analysis systems technique (F.A.S.T.) as a group knowledge elicitation method for model building." *Proc., 19th Int. Conf. Atlanta*, System Dynamics Society, Albany, NY.

SAVE International. (1998). "Function: Definition and analysis." <www.alyousefi.com/download/function.definition.analysis.doc> (Jan. 27, 2017).

Bytheway, C. W. (2005). "Genesis of FAST." *Value World*, **28**(2), 2–7.

Bytheway, C. W. (2007). *FAST creativity and innovation: Rapidly improving processes, product development, and solving complex problems*, J. Ross Publishing, Fort Lauderdale, FL.

Crow, K. (2002). "Value analysis and V system technique." <http://www.npd-solutions.com/va.html> (Jan. 26, 2017).

Johnson, P. (2013). "Fast diagramming made easy: Straightforward techniques for your highway project." <http://design.transportation.org/Documents/TC%20Value%20Engineering/2013%20VE%20Workshop/2013%20PPPs%20Papers_Wednes%20AM/1-Fast%20Diagramming%20Made%20Easy.pdf> (Jan. 27, 2017).

Kasi, M. (1994). "An introduction to value analysis and value engineering for architects, engineers, and builders: A continuing education course guide." <http://docshare01.docshare.tips/files/24027/240279184.pdf> (Jan. 28, 2017).

Kaufman, J. J. (2009). *Value management: Creating competitive advantage*, Sakura House, Etobicoke, Canada.

Lawrence, D. (2016) "Who was Larry Miles?" <http://www.valuefoundation.org/Innovate-VM-Miles.htm> (Jan. 26, 2017).

Miles, L. D. (1966). "The fundamentals of value engineering." <http://minds.wisconsin.edu/handle/1793/4434> (Jan. 25, 2017).

Miles, L. D. (1989). *Techniques of value analysis and engineering*, 3rd Ed., Lawrence D. Miles Value Foundation, Washington, DC.

SAVE International. (2007). "Value methodology standard and body of knowledge." <http://www.wsdot.wa.gov/nr/rdonlyres/34ffe1e3-bcc1-444d-93e4-d4dcf6ba3c3b/0/whatisve.pdf> (Jan. 25, 2017).

Stewart, R. B. (2010). *Value optimization for project and performance management*, Wiley, Hoboken, NJ.

Stocks, S. N., and Singh, A. (1999). "Studies on the impact of functional analysis concept design on reduction in change orders." *Constr. Manage. Econ.*, **17**(3), 251–267.

U.S. Department of the Interior, Bureau of Reclamation. (2008). "Reclamation value program handbook." <https://www.usbr.gov/dso-dec/vp/pdf/VPHandbook-7February2008.pdf> (Jan. 26, 2017).

Wixson, J. R. (1999). "Function analysis and decomposition using function analysis systems technique." <http://www.wvasolutions.com/622_p113.pdf> (Jan. 27, 2017).

CHAPTER 6

Decision Trees

6.1 Objective

This chapter aims to provide the reader with a sufficient understanding of what decision trees are and how they are implemented. Readers should understand the components in the structure of these diagrams and the proper notation for labeling them. They should understand the different types of situations where decision trees may be implemented and the topics they are most often used for, as well as the benefits and restrictions of this graphical tool, and be able to apply a decision tree to a problem of their choosing. After reviewing several example problems provided, readers should be prepared to complete the exercises found at the end of the chapter without guidance.

6.2 Introduction

Everyone faces multilayered problems with numerous possible solutions. The real problem, however, is deciding which outcome has the greatest worth to us. This worth can include profit, time, population, production, and other elements. One tool used to simplify these decisions is a decision-tree diagram. This is a graphical presentation of a decision problem that offers alternative sequential outcomes in a way that is easier for the user to visualize and follow. Through probabilistic and quantitative analysis, we can make a logical decision about a problem with the aid of a decision tree. This tool is particularly helpful for choosing between different strategies with potentially positive outcomes. The diagram is generally drawn from original decision to final effect, based on all known data, until all considered results are represented. The final product appears as a sideways connection of branches— usually from left to right—that vaguely resembles the clever name given to it: a tree of decisions.

6.2.1 Understanding the Potential and Limits of Decision Trees

Decision trees have become increasingly helpful in the fields of business, marketing, accounting, science, and engineering, which all present quantitatively complex problems with many potentially different solutions. There is almost always some form of probability and revenue associated with decision-tree problems, which is why professionals in these fields turn to their aid. However, it is important to understand that decision trees are not a panacea—some magical formula that will automatically solve any problem relating to statistical data and finances. To successfully implement this tool, one must have enough data and an intuitive understanding of the field. A decision tree is only as accurate as the data provided in the problem statement. When applying this method to a real-world problem, confidence in the numerical data provided is essential for an accurate analysis. Before we dive too deeply into the qualitative structure of a decision tree, it may be helpful to expose ourselves to a very simple example. Take, for instance, the following scenario involving a marketing problem.

6.2.1.1 Marketing Problem

The marketing department of Engineered Shoos, Ltd. has filmed a commercial for a new athletic shoe. The cost of filming the ad was $15,000, and it has been edited to 20 seconds in length. Engineered Shoos is planning to feature the commercial on ESPN, during its prime-time hours, a total of 100 times over three weeks. During regular prime-time hours with no special events, ESPN charges $1,000 per second for advertising time. Through many years of statistical analysis, Engineered Shoos has deduced that when television advertisements are used in 100-block increments as proposed, there is a 60% chance that profits from this new shoe will increase by $8 million and a 40% chance that they will increase by $2 million. The question is whether this advertising strategy is financially smart. A decision is required.

This problem can be solved without drawing a decision tree, but with all the quantitative data involved, it might be easier for us to analyze the problem with a graphical aid. Refer to Fig. 6-1.

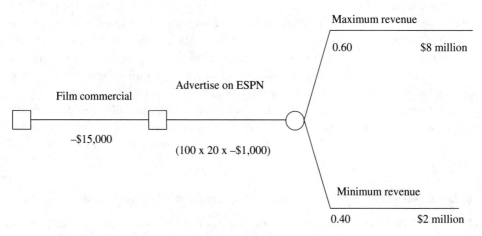

Fig. 6-1. Projected outcomes from the Engineered Shoos Ltd. television commercial

The simplest way to solve this problem visually is to follow every possible path and calculate the result. This problem happens to deal with money, and we are interested in the end profit that is calculated at the far right side of the diagram. Step by step, we can follow each series of events and numerically compute the whole chain. There are only two possible paths:

Film commercial → Advertise on ESPN → Maximum revenue →

$$(-\$15{,}000) + (100 \times 20 \times -\$1{,}000) + (0.60 \times \$8{,}000{,}000) = \$2{,}785{,}000$$

Film commercial → Advertise on ESPN → Minimum revenue →

$$(-\$15{,}000) + (100 \times 20 \times -\$1{,}000) + (0.40 \times \$2{,}000{,}000) = -\$1{,}215{,}000$$

Intuitively, we can see that our two different paths are both potential outcomes of a decision made at an earlier stage. Since they share this common history, the expected outcomes need to be added to determine the probabilistic outcome for the original decision.

$$\text{Total probable profit} = \$2{,}785{,}000 + (-\$1{,}215{,}000) = \$1{,}570{,}000$$

Now we must ask ourselves if an expected profit of $1,570,000 is financially smart, considering all the time and effort that would be invested in planning for and creating this advertisement. And, does the advertising department feel that this is worthwhile or of adequate efficiency? This is where an intuitive understanding of the field comes in. Had we projected a total profit of $500 yet had no understanding of marketing, we might have said, yes, this campaign seems financially smart, because the profit is greater than $0. This leaves a question of scale; about $1.57 million potentially not being considered financially efficient enough for Engineered Shoos.

Following that very basic example should have made the positive and negative aspects of using decision trees clearer. Maybe a quick discussion of the positive and negative things we noticed about this graphical tool will help.

Positives:

- The decision tree helps to structure a problem into a graphical model so that each aspect can be broken down individually and then compared with other possible decision paths.
- When there is a significant amount of information given that relates to multiple aspects of a problem, it can be confusing to try to apply the data directly to a formula. Instead, we can configure the problem visually so that we know which events precede others and how probabilities affect decisions made further down the line.

Negatives:

- We must first have ample prior statistical data about each potential decision or outcome path, which often requires extensive research. The data given in the Engineered Shoos problem—that maximum revenue had a 60% probability—was not chosen randomly. However, this is a limitation of the entire expected-value analysis theory, not just of decision trees.
- A completed decision tree has many complex and detailed components, so it is important to have a reliable and accurate source of data to ensure that our decision tree yields useful results.
- Decision trees can be time-consuming to create, which can be inconvenient when facing tight deadlines.

6.3 Diagram Breakdown

The decision-tree diagram always begins with the base node, or root. This reference spot represents the point from which all paths branch out. Because we cannot begin solving a problem without some initial decisions, we call this base node a *decision node* and represent it with a square box. Whenever there is a decision node, the branches that stem out must represent decisions made by the individual solving the problem. These are represented as straight lines, with the name of the decision written above the line and the pertinent quantitative information written below.

Typically, there are at least two decisions branching out of every decision node because usually every choice has an alternative. However, it is not always necessary to display every option. In the example given in Section 6.2, only one decision was made for the first two nodes that appear in the tree. An alternative could have been to film a different commercial and televise it on several stations with differing air-time costs. This choice was not included in the example, which nevertheless still fulfilled the requirements for a decision-tree diagram.

A decision branch almost always has a number associated with it, which we display below it at the far right side. This number can represent various interests—e.g., money, population, or time. Typically, no probability is connected to a decision branch because this is our degree of freedom, and probabilities are related to uncertain outcomes. However, the uses of decision-tree diagrams are nearly endless, and they are too generic to try to classify so specifically. Fig. 6-2 shows the beginning branches of a rudimentary decision tree.

From here, each branch can end, lead into a decision node, or lead into a *chance node*. Converging into a decision node would yield the same situation as before. Chance nodes are breaks in the path where one no longer has a choice in making a decision. They are different from decision nodes and are therefore represented with a circular box. These nodes are typically followed by more than one outcome, displaying possible results—based on probabilistic data—of the previous decision. Outcome

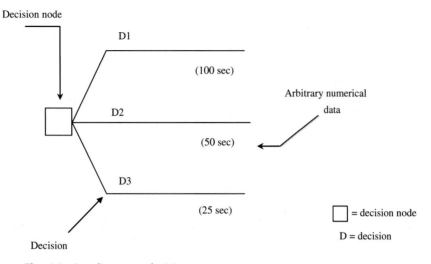

Fig. 6-2. A rudimentary decision tree

branches look much the same as decision branches, except that they stem from chance nodes and not decision nodes. To accurately analyze the problem statement, each outcome must be linked to a probability. Because we are not certain of the direction the problem will take through a chance node, a probability is the only tool we have for weighing the seriousness of that path. Obtaining accurate statistical data for these outcome paths is probably the most difficult and time-consuming task in preparing the problem statement for decision-tree analysis. Fig. 6-3 is an example of what the second-tier branches of a decision tree might look like.

This process is repeated until each aspect of the problem is mapped and all quantitative data are displayed. Many decision trees are quite intricate and detailed, as shown in Fig. 6-4.

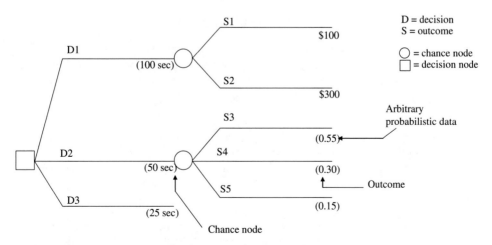

Fig. 6-3. Expansion of rudimentary decision tree

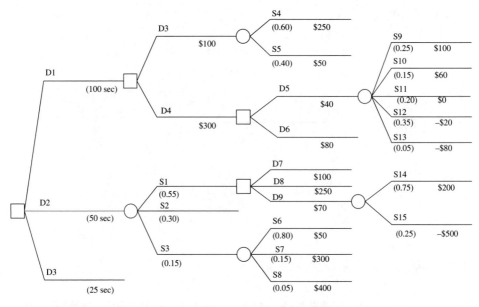

Fig. 6-4. Further expansion of the rudimentary decision tree

6.4 Example Problems

Now that we have familiarized ourselves with the structuring of a decision tree and proper notation, let us take a look at some examples. Example problems are the best way to learn how to use this hands-on tool. These three potentially probabilistic scenarios offer a detailed guide to the beneficial use of decision-tree diagrams. Explaining the breakdown of a problem through a step-by-step analysis will allow us to see the careful process by which a decision tree is created and help us become more independent when applying this tool to our own problems.

6.4.1 Example 1: Application in Construction Engineering

A very common engineering decision is whether to bid on a certain construction project. Some projects offer high levels of risk with high payouts, and others offer lower levels of risk with a smaller profit. Many times, specific individuals are hired by these engineering firms to decide which projects to bid on and which to pass on. In a busy construction market, there are always more jobs available than any company can handle, and companies often find themselves unable to bid on every job.

For example, ABC Engineering is considering bidding on two different construction projects around Honolulu. The first project is a four-story parking garage; the estimated cost of bidding on the contract is $10,000[1]. On the basis of the

---

[1]For those who haven't had the experience, cost estimation on a $3 billion coal power plant can take six months, require a large team of estimators and support staff, and cost close to a million dollars in salaries, travel, production of schedules and drawings, and other expenses. Hence, the decision whether to spend that much time and money is very important.

firm's history of risk-taking and competitiveness for the project, it has been estimated that the firm has a 75% chance of losing the bid and a 25% chance of winning. Furthermore, if ABC's bid is awarded, ABC has a 25% chance of earning $250,000, a 40% chance of earning $150,000, a 20% chance of earning $80,000, a 5% chance of breaking even (earning $0), and a 10% chance of losing $20,000. The second project is for a new high school; the estimated cost of bidding on this contract is $8,000. Through experience, they have determined that they have a 65% chance of losing the bid and a 35% chance of winning. Earning probabilities come in at a 45% chance of earning $300,000, a 30% chance of earning $210,000, a 10% chance of breaking even, and a 15% chance of losing $15,000. Is it financially smart to bid on either of these contracts? And if ABC were to have only enough resources to bid on one project at a time, on which one should the firm spend its time?

Step 1: We need to start at the base node and determine which decision path we take from there. The first sentence of the problem states that there is the potential of bidding on two different projects. So, either the firm can bid on the parking garage or the high school. (It is implicitly understood that a possible third decision is not to bid on either project, but it is not necessary to include that option in the decision tree.) We must also remember that there is a cost associated with each of these decisions that must be displayed on the decision tree as well. It costs $10,000 to bid on the parking garage and $8,000 to bid on the high school project. The beginning node and decision branches for this problem should look like Fig. 6-5.

Step 2: For each of these branches, we need to decide whether it leads into a decision node or a chance node. There is a chance that ABC can either win or lose the bid on either project. The word "chance" directly suggests that the two bid decisions will be followed by chance nodes. However, even if the problem statement had not used the exact word "chance," we would have still known this to be true, because winning or losing the bid is not within ABC's control. The firm does not make the decision to win or lose; rather, there is some probabilistic possibility of either outcome happening.

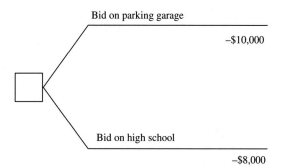

Fig. 6-5. Basic decision path for ABC Engineering

Step 3: The different possible outcomes for both of these chance nodes need to be decided. For the parking garage project, ABC has a 75% chance of losing the bid and a 25% chance of winning it. For the high school, ABC has a 65% chance of losing the bid and a 35% chance of winning it. Losing the bid in either case yields a net gain of $0, so we are free to write that value below both decision branches. Now that we know our different outcome paths and probabilities, we can draw something similar to Fig. 6-6.

Step 4: Now we must decide whether our outcome paths lead into decision nodes or chance nodes. Intuitively, it is clear that if ABC loses a bid, then that is the end of that path. But if ABC wins the bid on the parking garage, then there are five different possible outcomes, none of which are directly within the firm's control. Likewise, for the high school, there are four different possible outcomes, none of which are directly within ABC's control. Both of the decision paths stemming from "Win bid" therefore lead into chance nodes.

Step 5: From both of these chance nodes, the different possible outcomes need to be determined. For the parking garage project, ABC has a 25% chance of earning $250,000, a 40% chance of earning $150,000, a 20% chance of earning $80,000, a 5% chance of breaking even at $0, and a 10% chance of losing $20,000. The five different outcomes have probabilities and earnings associated with them. For the high school project, ABC has a 45% chance of earning $300,000, a 30% chance of earning $210,000, a 10% chance of breaking even at $0, and a 15% chance of losing $15,000. These four different outcomes also have probabilities and earnings associated with them. Our decision tree should now look similar to Fig. 6-7.

Step 6: There are no more data to cover, so we have concluded the graphing part of the decision tree. We now must calculate the potential profit or loss from following any individual path on the tree. For an example of a path, see Fig. 6-8.

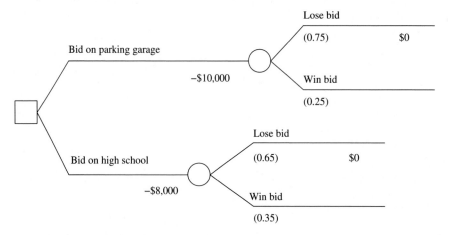

Fig. 6-6. Expanded decision tree for ABC Engineering

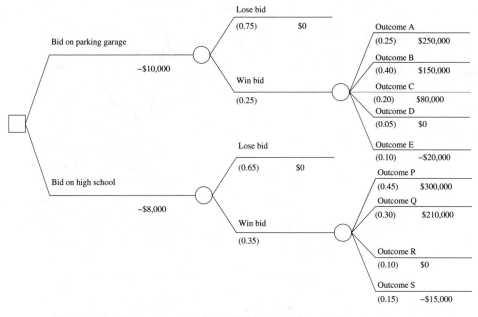

Fig. 6-7. Further expansion of the ABC Engineering decision tree

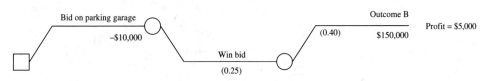

Fig. 6-8. An individual path from the ABC Engineering decision tree

If we follow the path in Fig. 6-8, we can calculate the resulting profit or loss with the following equation:

$$Profit = -\$10{,}000 + [0.25 \times (0.40 \times \$150{,}000)] = \$5{,}000$$

The first step is bidding on the parking garage; this is ABC's choice, and there is no probability associated with it, which is why we start by subtracting $10,000. Next, we assume that ABC wins the bid, but because this is an outcome branch there is a probability of 25% associated with it. Because there is no monetary value related to that specific branch, the value of any branch that stems from it must be multiplied by this number. Finally, we must multiply the $150,000 profit by ABC's 40% chance of earning it to determine the statistical revenue for that outcome.

Following this procedure, we can then calculate the profit or loss for each individual path and display it on the decision tree, as in Fig. 6-9.

Step 7: The last step is to sum all the profit and loss that occurs in the final outcomes that stem from the chance nodes to determine the total expected profit or loss for the project. The decision tree should now look similar to Fig. 6-10.

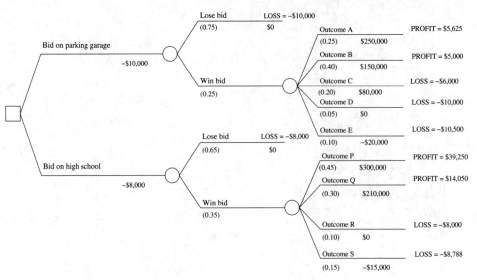

Fig. 6-9. Profit and loss calculations for the ABC Engineering decision tree

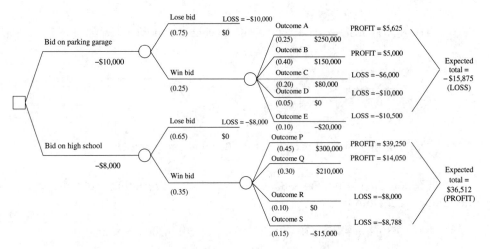

Fig. 6-10. Total profit and loss for ABC Engineering

This procedure requires some intuitive thinking, and in the end, we must make a comparison that decides the choice that the contractor should make. It is clear that, probabilistically, ABC will lose money by bidding on the parking garage project but will make money by bidding on the high school project.

Answer: It is financially smart for ABC to bid on the high school but not on the parking structure.

6.4.2 Example 2: Application in Fundraising

After going through the previous example in great detail, readers should have a better understanding of how decision trees should be set up. For the following example, there will be less time spent on explaining concepts but an equal amount of time spent on displaying the appropriate decision-tree graph. A fundraiser is a perfect application for this tool because it deals with profit and the statistical data are, practically speaking, feasible to obtain.

Assume that the Hawaiian student chapter of the American Society of Humanitarian Engineers (ASHE) is thinking of ways to raise money and decides to hold the "Tour d'Engineer" beer-drinking festival and charge $15 per person. The organizers decide to open it to the entire school in hopes of a larger turnout that would also yield a greater profit. These crafty students recently took an undergraduate statistics course, so they decided to perform some careful statistical analysis of the different possible dates on which they could hold the event.

The first possible date is April 3. The students have discovered that the price of beer and entertainment on that day will result in an initial cost of $300. They have also discovered that the weather report predicts a 35% chance of rain that day and a 65% chance of clear skies. If there is no rain, the students may hold the event outside in the amphitheater without cost. However, if rain persists, the students need to decide whether they should remain outside uncovered at no additional cost, rent tents for $50, or move the party inside at a cost of $100. If there is no rain and the students remain outside, they predict a 70% chance of a high turnout of 425 attendees and a 30% chance of a low turnout of 255 attendees. They also predict that they will see a 40% decrease in attendance if it rains and they rent tents, an 80% decrease if it rains and they stay outside without cover, and a 10% decrease if it rains and they move the event inside. The decrease in attendance applies to both the high-turnout and low-turnout predictions.

The second possible date is April 10. Beer and entertainment on that day will cost them an initial amount of $450. The weather report predicts a 20% chance of rain and an 80% chance of clear skies. If there is no rain, the students may hold the event outside in the amphitheater for free. However, if rain persists, the students need to decide whether they should remain outside uncovered at no additional cost, rent tents for $50, or move the party inside at a cost of $100. If there is no rain and the students remain outside, they predict a 70% chance of a high turnout of 380 attendees and a 30% chance of a low turnout of 228 attendees. As on the previous date, the students predict that they will see a 40% decrease in attendance if it rains and they rent tents, an 80% decrease if it rains and they stay outside without cover, and a 10% decrease if it rains and they move the event inside. The decrease in attendance is assumed for both the high-turnout and low-turnout predictions. On which date is it more financially beneficial to hold the event?

Step 1: From the base node, the students have a decision to make: They hold the event either on April 3—which would have an initial beer and entertainment cost of $300—or on April 10—which would have an initial beer and entertainment cost of $450. (See Fig. 6-11.)

Step 2: On either of the two potential dates, there are unknown weather conditions that are out of the students' control. Both of the following junctures

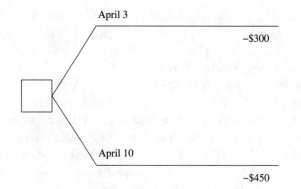

Fig. 6-11. Basic decision tree for the ASHE Beer Festival

must thus be chance nodes. The outcomes that can occur on April 3 are rain at a 35% probability and no rain at a 65% probability. On April 10, we have rain at a 20% probability and no rain at an 80% probability. Fig. 6-12 adds this information to the decision tree.

Step 3: If rain occurs, it is followed by a decision node because the students can (1) rent tents for $50, (2) stay outside uncovered at no charge, or (3) move everyone inside for $100. If there is no rain, they have decided to stay outside at no extra cost. These options apply to both the April 3 path and the April 10 path, as shown in Fig. 6-13.

Step 4: Since ASHE has no control over the turnout of attendees to the event, we know that each decision branch leads into a chance node. From each chance node stems either a high-turnout outcome, with a 70% probability, or a low-turnout outcome, with a 30% probability. However, each outcome branch must include the possible reduction in attendees resulting from the weather and the decision on amenities in case of rain. A sample calculation for a randomly chosen case is:

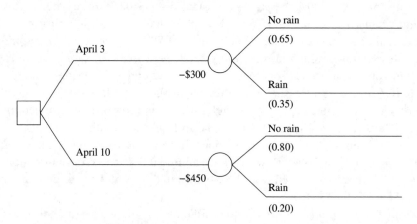

Fig. 6-12. Expanded decision tree for the ASHE Beer Festival

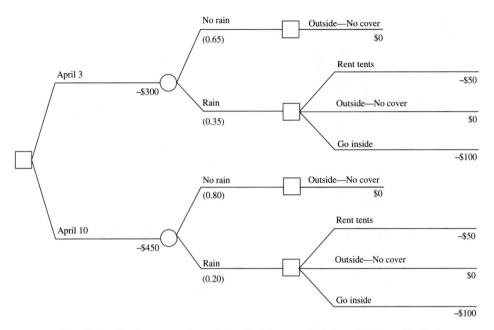

Fig. 6-13. Further expansion of the decision tree for the ASHE Beer Festival

April 3 → Rain → Tent rental → High turnout $[425 - (425 \times 0.40)] \times \$15$

Because a 40% reduction in attendance is applied to a high turnout of 425 people, we multiply 425 by 0.40 to obtain the number of participants who will bow out of the event owing to the weather. We subtract that from the original 425 who would show up if there were no rain and obtain the total number of people who will attend after the reduction is calculated. Then we multiply that total by $15, the price per person that ASHE will be charging. This price must also be multiplied by 0.70, because that is the probability of a high turnout. These calculations are shown on the diagram in Fig. 6-14 for easier reference.

Step 5: Calculate the profit or loss for all decision paths that end at the outcomes on the far right side of the diagram. Fig. 6-15 shows an example path from the diagram.

If we follow this example path, we can calculate the resulting profit using the following equation:

$$Profit = -\$300 + [0.35 \times (\{0.70 \times [425 - (425 \times 0.10)] \times \$15\} - \$100)] = +\$1,071$$

Following this procedure, we can then calculate the profit/loss for each individual path. Sum up all of the ending outcomes that relate to April 3, and all of the ending outcomes that relate to April 10. We can display the finalized decision tree in Fig. 6-16.

Answer: It would be more financially beneficial to hold the beer festival on April 3. Thus, this student engineering society has a viable path to increased revenue generation. The decision is based on well-considered objective data.

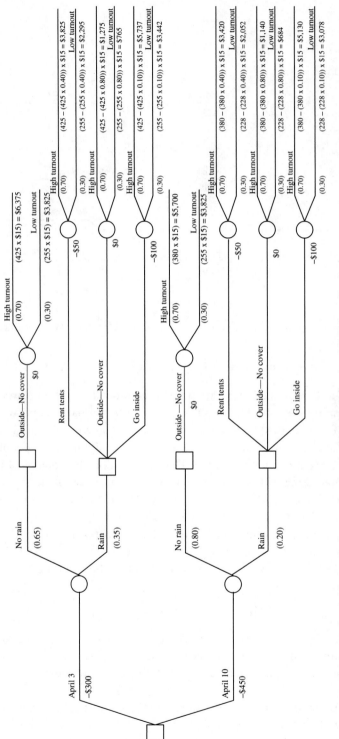

Fig. 6-14. Profit and loss calculations for the decision tree for the ASHE Beer Festival

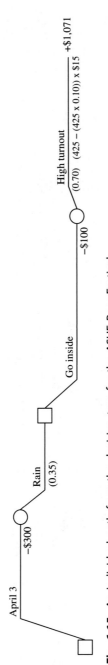

Fig. 6-15. An individual path from the decision tree for the ASHE Beer Festival

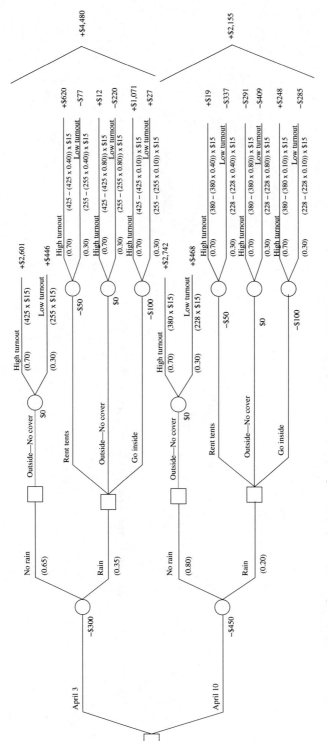

Fig. 6-16. Total profit and loss calculations for the decision tree for the ASHE Beer Festival

6.4.3 Example 3: Application in Insurance

Here is an example of a problem generally not thought of as appropriate for the use of a decision tree; however, through a perceptive understanding of the problem statement, we can adjust it to work with our diagram tool. This will be more difficult than the previous two examples, because careful attention must be directed to the placement of the decision tree's components.

Consider a car insurance company and its desire to further examine the breakdown of profit within its client base. This is very pragmatic, because a systematically run organization such as a car insurance company will have volumes of stored statistical data on individual client records.

Rotunda Car Insurance has been experiencing a slight decline in profit for the last few years and decided to analyze its client base to see if there are any categories that could generate additional revenue. Rotunda organizes its database information into two basic demographics: men and women. From there, the clients are broken down into two different age categories: 16 to 40, and 41 and older.

The firm has decided to take a blind sample poll of its client base, using equal numbers of randomly assigned surveys in each of the two age categories among both men and women. The data from these random trials can be further broken down into individuals who have had no more than one car accident (that is, zero or one) in the past two years and individuals who have had two or more accidents in the past two years. Furthermore, the company has probabilistically calculated the chances that an individual from each category will have an accident in the next two years.

The individuals in each age group pay different premium rates based on the number of accidents that they have had in the past two years. In the category of men ages 16 to 40 with at most one accident, 321 individuals will each pay $1,680 over the next two years and have a 17% chance of being involved in an accident within the next two years. The 12 men ages 16 to 40 with two or more accidents will pay $2,160 and have a 69% chance of an accident. In the category of men ages 41 and over with at most one accident, 309 individuals will each pay $1,824 over the next two years and have a 19% chance of an accident. The 24 men ages 41 and over with two or more accidents will pay $2,304 and have a 72% chance of an accident.

In the category of women ages 16 to 40 with zero or one accident, 315 individuals will pay $1,752 over the next two years and have a 19% chance of having an accident in that time. The 18 women ages 16 to 40 with two or more accidents will pay $2,232 in insurance and have a 72% chance of an accident. The 302 women ages 41 and over with no more than one accident will pay $1,896 over the next two years and have a 23% chance of an accident, and the 31 women ages 41 and over with two or more accidents will pay $2,376 and have a 78% chance of an accident.

Rotunda has further estimated that each accident claim costs the firm an average of $9,000, because some accidents involve expensive new cars that are totaled. Which client base—men or women—produces less revenue for Rotunda? More specifically, which age demographic in that group is the costliest, and which demographic earns the company the most money?

Step 1: Since we are analyzing demographic groups starting from the most general and working down to more specific categories, it would be sensible to start with men and women as separate decision paths from our base node (see Fig. 6-17). What makes this problem different from those discussed previously is that there are no numerical data provided for these two largest groups. Nevertheless, choosing which path to follow in an analysis is a decision within our control.

Step 2: The next logical path would be breaking down each gender into more specific age categories. Because we may choose either path, and because Rotunda has chosen to categorize its data in this way, this juncture is a decision node. Again, there have been no quantitative data provided for either of these decisions, but we can still apply it to our decision tree model, as seen in Fig. 6-18.

Step 3: Each age group is then divided into two different classifications as determined by Rotunda: individuals who have had zero or one accident in the past two years and individuals who have had two or more. We have been given data in each of these categories for the number of people fitting these criteria and the total

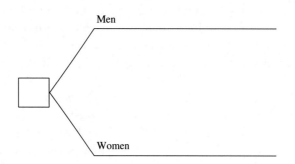

Fig. 6-17. Basic decision tree for the insurance case

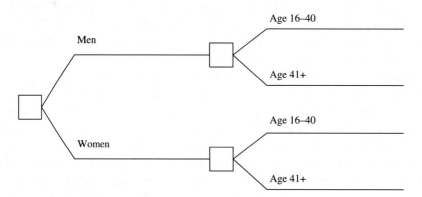

Fig. 6-18. Expanded decision tree for the insurance case

premiums charged to members of each category over a two-year span. By multiplying the population by the premiums, we can obtain the total revenue received by Rotunda from that demographic over the two-year period. Again, because we may choose to trace either path, the age groups lead into decision nodes. This information is shown in Fig. 6-19.

Step 4: The last piece of information given in the problem statement is the probability of an individual member of each group being involved in an accident during the next two years. Only the data for the chance of an accident were provided, but we can easily find the numerical values for the probability of *not* being in an accident. We were also told that the average accident claim costs Rotunda $9,000; it is implicit that the cost of a client who does not have an accident is $0. The trick is realizing that we must multiply the average accident cost by the probability of an accident and then by the population of individuals in that category (see Fig. 6-20).

Step 5: Calculate the profit or loss for all decision paths in the diagram. See the path shown in Fig. 6-21, we can calculate the profit as

$$Profit = (309 \times \$1{,}824) + (0.81 \times 309 \times \$0) = +\$563{,}616$$

Following the same procedure, we can then calculate the profit or loss for each individual path. Add up all of the ending outcomes for each specific age group for men and women, then do the same for men and women as a whole. The final decision tree is shown in Fig. 6-22.

Answer: Overall, men bring more revenue to the company than women. This result emerges from the facts of the problem and not because of any gender bias.

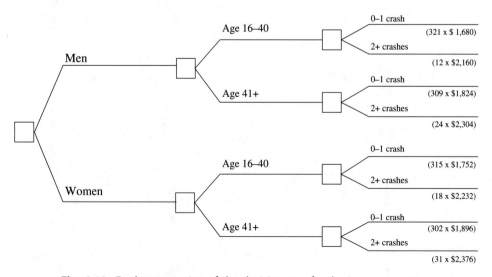

Fig. 6-19. Further expansion of the decision tree for the insurance case

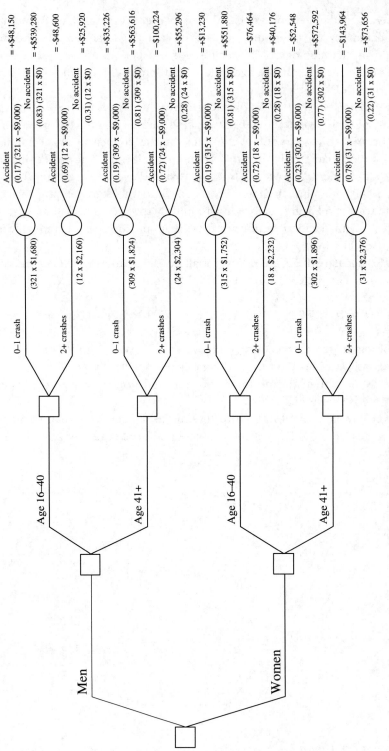

Fig. 6-20. Cost calculations for the decision tree for the insurance case

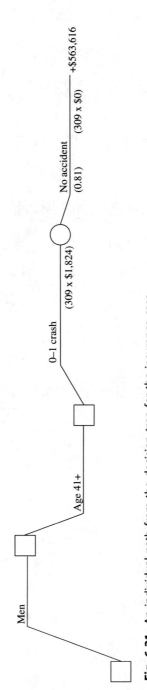

Fig. 6-21. An individual path from the decision tree for the insurance case

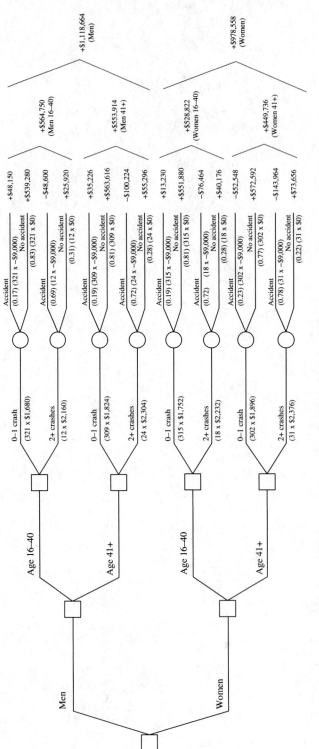

Fig. 6-22. Total profit and loss calculation for the decision tree for the insurance case

Women ages 41 and over bring in the least revenue, and men ages 16 and 40 bring in the most. This information should help the insurance company in formulating its policies and strategies and in deciding the premiums it should charge individual clients based on the different risks they pose. In fact, insurance companies take many factors into consideration when setting premiums, including age, accident record, driving experience, and so forth.

6.5 Review

As we close this chapter, readers should now have an understanding of what a decision tree is and the function it performs. Having covered its structured components in detail, readers should know the proper notation and labeling for this handy tool, as well as that fields such as business, marketing, accounting, science, and engineering especially benefit from the use of these diagrams because of their use of quantitative data in analysis problems. Moreover, through the three examples worked through in section 6.4, it should also be clear that the specific uses of decision trees can stretch far beyond our means of direct classification. With this understanding of the proper steps for implementing a decision tree for a given problem statement, readers can further test their knowledge of the field by working through the following exercises.

6.6 Exercises

Problem 1

Grand Construction Co. is debating placing a bid on a construction project for a new city library in downtown Honolulu. However, the two partners of the firm have conflicting views on the level of confidence they should have before placing a bid. The first partner likes to be safe and take a low level of risk in order to avoid losing money on the final project; the other partner prefers to take greater risks so that the firm has a better chance of being awarded the contract. The estimated cost of bidding on the project, regardless of risk, is $9,000. If the risk-averse partner has his way, the firm has a 20% chance of winning the contract and an 80% chance of losing it; if the risk-tolerant partner has his way, the firm has a 65% chance of winning the contract and a 35% chance of losing it. If the contract is awarded under low-risk conditions, the firm will have an 80% probability of earning a $150,000 profit, a 10% probability of breaking even at $0, and a 10% probability of losing $50,000; under high-risk conditions, the firm faces a 50% probability of earning a $150,000 profit, a 25% probability of breaking even at $0, and a 25% probability of losing $50,000. Is it financially smarter for Grand Construction to bid on the contract under low-risk or high-risk conditions?

Problem 2

A young professional is debating whether she should start her own office-supply sales business or stay with the sales job she currently has. She wants to be fully aware of the possible consequences before making a decision and has chosen to analyze her total profit over the next four years under each possible scenario. She has estimated that initial startup costs for the business would be $30,000, as opposed to $0 if she stays at her sales job. She is aware that, in general, new small businesses have a 60% chance of failing at some point in their first four years; she wants to compare this to only a 10% chance of being fired from her job within the next four years. If she succeeds at her new endeavor, she is confident that she has a 65% probability of bringing in a maximum of $600,000 in revenue over the first two years and $750,000 over the third and fourth years, compared to a 35% probability of bringing in a minimum of $80,000 over the first two years and $100,000 over the third and fourth years. However, if she keeps her current job and is not fired, she believes that she has a 75% probability of earning a maximum of $200,000 during the first two years and the same again during the next two years, compared to a 25% probability of earning a minimum of $150,000 over both the first two years and the last two years. Which business decision would be more financially prudent?

CHAPTER 7

Payoff Matrix

7.1 Introduction

Payoff matrices serve as important tools in risk analysis and decision making. A payoff matrix can be used to identify risk in both every-day and multibillion-dollar business decisions. Payoff matrices are mostly used to determine the potential profit from or cost of a decision and the corresponding risk, as well as to make qualitative decisions in which the decision maker compares alternatives with desired characteristics. This tool enables the decision maker to make an informed decision by use of four different strategies: finding the *maximin, maximax, minimax regret,* and *expected monetary value* of each situation. The *maximin* is the best of the worst-case scenarios; the *maximax* is the best of the best-case scenarios; the *minimax regret* is used to decrease opportunity loss; and the *expected monetary value* is the expected value based on the probability of each occurrence.

The payoff matrix method breaks the decision process down into *decision alternatives* and *states of nature.* The decision alternatives of a problem are all the decision maker's possible decisions, whereas the states of nature are the possible actions that can be taken to address the decision alternatives. The decision maker may be able to determine the probability of occurrence for each state of nature and choose a desirable path accordingly.

7.2 Method

The first step of the payoff matrix process is to determine the decision alternatives—i.e., the possible decisions. There may be an unlimited number of decision alternatives, but there must be at least two, or else there will be no decision to make. After the decision alternatives have been determined, we must determine the states of nature. There are two different types of states of nature, which depend on what type of problem is being solved: If it is a quantitative problem dealing with profit or cost, then the states of nature are the economic reactions to the decision; if it is a qualitative problem, then the states of nature are whatever the decision maker feels are important qualities to weigh.

The next step is to analyze each decision alternative and assign it a value for each state of nature. If the payoff matrix is being used to determine profit or cost, then the likely profit or cost of each decision alternative under a certain state of nature should be determined.

A qualitative payoff matrix is done slightly differently. In this situation, each decision alternative is graded for each state of nature. After each decision alternative has been weighed against each state of nature, a table is constructed with the decision alternatives in the left-hand column, the states of nature across the top row, and the corresponding weights filling out the table. Table 7-1 shows a basic example.

Once the information has been gathered and the table has been constructed, the decision maker can then determine the maximin, maximax, minimax regret, and expected monetary value. These methods may yield different recommendations, but then it is up to the decision maker to decide which method to give preference to, depending on propensity for or aversion to risk. When the payoff matrix is done to determine costs, the answers are determined similarly.

The *maximin* is short for the maximum of the minimum values. Decision makers would be risk averse if they were to make their decisions based solely on the maximins. To determine the maximin, we must first determine the minimum values of all decision alternatives. In some problems, the minimums will all be associated with the same state of nature, but this is not always the case. Once all of the minimums have been determined, the decision maker must choose the highest of all of these minimums, or the maximin.

Another decision-making method is to use the *maximax*, which represents the maximum of the maximum values. The probability of this value occurring does not matter; the decision maker is simply looking for the best possible outcome. The maximax can be determined simply by finding the highest number on the table.

The third strategy used with the payoff matrix is the *minimax regret*. This method is useful in limiting the regret that the decision maker would feel under any potential state of nature. It can also be understood as the minimum of the maximum opportunity loss. To determine the minimax regret, the *regret* (opportunity loss) for each decision alternative—the difference between the value of the given decision and the maximum possible value—must be determined for each state of nature. For example, if decision alternative 1 (D1) received a score of 5 for state of nature 2 (S2), and the maximum score under S2 is 9, then the level of regret for that decision is $9 - 5 = 4$. Levels of regret must be determined for each decision alternative in each state of nature, and the

Table 7-1. Example payoff matrix table

	State of nature 1 (S1)	State of nature 2 (S2)
Decision alternative 1 (D1)	5	8
Decision alternative 2 (D2)	6	12
Decision alternative 3 (D3)	7	9

maximum level of regret must then be determined for each decision alternative. The minimax regret is the minimum of all of the maximum regrets. This becomes clearer through the three example problems examined later in this chapter.

The final and perhaps most informative step in the payoff matrix is to determine the *expected monetary value* (EMV). This is the only step in the payoff matrix that includes all possible states of nature in its final answer. The decision maker must also have additional information about the probability of occurrence for each state of nature. For a quantitative payoff matrix, these probabilities should be based on research; for a qualitative payoff matrix, they may be subjectively determined. The probability of each state of nature is multiplied by the corresponding value for that decision alternative, and the sum of the results is the EMV of that decision alternative. For example, a decision alternative is expected to turn a profit of $20,000, $10,000, or $500, with the probabilities of the corresponding states of nature being 0.5, 0.4, and 0.1 respectively[1]. The EMV can be calculated as ($20,000 × 0.5) + ($10,000 × 0.4) + ($500 × 0.1) = $14,050. This value can then be compared to the EMVs of all other decision alternatives; the decision alternative with the highest EMV, considering all states of natures and their chances of occurrence, is the best choice. EMV can be a helpful tool for solving a problem when the decision maker knows how to use it properly.

Decision trees can also be used when determining EMV, because they follow the same steps. Decision trees were explained in Chapter 6, but Example Problem 1 in the following section includes a short rundown of how to use a decision tree to determine EMV.

Risk comes into play when deciding which method to use to make the ultimate decision. Are the decision makers a risk taking or risk averse? If decision makers are the gambling type, they might choose the maximax, hoping for the best-case scenario[1]. If the decision makers are risk averse, they would probably choose the maximin or the minimax regret, both to reduce the chance of loss. Important decisions for which the probabilities of occurrence are known should be made using EMV or a similar method.

The following three example problems will help the reader understand the steps of the payoff matrix. The problems highlight the differences in payoff matrices used to determine profits, costs, and qualitative results. The first example problem demonstrates how to use a payoff matrix to determine the best choice for maximizing profits.

7.3 Example Problem 1: Increasing Monthly Profits

A small environmental engineering company wants to increase its monthly profits. Currently, half of the firm's projects are green buildings, and the other half are

---

[1]The sum of the probabilities must be 1.0.

research-based; its monthly profit is $30,000. The firm's owner is contemplating changing the company's focus to either only research or only low-impact development (LID) projects.

The owner has determined that two scenarios could ensue from changing the company's business strategy given growing environmental concern in the community. Either citizens, and the government that represents them, may become more environmentally friendly, or the corporations that plan to build in the area may become more environmentally aware.

The owner has determined that if the government and citizens become more environmentally conscious, it would be best to switch to focus on research in order to receive more government funding. The expected profits would rise to $35,000 per month under this scenario. If the firm became solely a low-impact development company, profits would actually decrease to $28,000 per month, given these circumstances.

If the corporations in the area decide to become more environmentally friendly, then focusing on low-impact development would increase profits to $34,000 per month. In these circumstances, the company's profits would fall to $20,000 per month if it chose to solely conduct research. If the company does not change its current business plan, the average monthly profit will remain $30,000 per month regardless of conditions. The owner has also determined that there is an equal chance that each market reaction will occur.

Given this information, determine the maximin, maximax, minimax regret, expected monetary value, and expected opportunity loss of the engineering company's three decision alternatives.

7.4 Answer to Example Problem 1

7.4.1 Determining Decision Alternatives

The first step is to determine the decision alternatives for the problem. In this case, they are the owner's three possible decisions. Decision 1 (D1) is to keep the current business strategy of 50% research and 50% low-impact development; decision 2 (D2) is to conduct only environmental research; and decision 3 (D3) is to work only on low-impact development projects.

7.4.2 Determining States of Nature

The second step is to determine the states of nature for the given problem, or the possible results of the decision alternatives. The first state of nature (S1) is that the citizens and their government will be swayed by the company to become environmentally friendly ("green" citizens and government). The second possible state of nature (S2) is that the corporations in the area will become more environmentally friendly (green companies).

Once the decision alternatives and the states of nature are determined, the information should be put into a table. The states of nature are placed along the top of the table and the decision alternatives in the far left column of the table. The monthly profits are the results of comparing each decision alternative under each state of nature. The table for this problem should be filled out as shown in Table 7-2.

7.4.3 Determining Maximin

The first part of the question asks us to determine the maximin for this scenario. To do this, the minimum profit of each decision alternative must first be determined. The minimum profit for D1, D2, and D3 are $30,000, $20,000, and $28,000, respectively. The maximin is simply the highest of all of these minimums, which is $30,000 for decision alternative 1. This is shown in Table 7-3, where the maximin is circled.

7.4.4 Determining Maximax

The second part of the question asks for the maximax of the scenario. This is determined by using the maximum possible profit for each decision alternative under each state of nature. Again, the profits from each decision alternative must be examined and the highest possible profit determined. For D1, not changing the company's strategy, the highest possible profit is $30,000 per month. D2 results in a maximum profit of $35,000 given the state of nature of environmentally friendly citizens and government. D3 has a maximum profit of $34,000 per month given the state of nature of environmentally friendly corporations. Therefore, the maximax of this problem is the $35,000 profit that results from D2 given the state of nature of green citizens and government. This is shown in Table 7-4, where the maximax is circled.

Table 7-2. Payoff Table: Business Strategy vs. Expected Monthly Profit

Decision alternative	Green citizens and government (S1)	Green companies (S2)
Half research and half LID (D1)	$30,000	$30,000
Research (D2)	$35,000	$20,000
Low-impact development (D3)	$28,000	$34,000

Table 7-3. Maximin of Research vs. Green Building Scenario

Decision alternative	Minimum payoff
Half research and half LID (D1)	$30,000 ← Maximin
Research (D2)	$20,000
Low-impact development (D3)	$28,000

Table 7-4. Maximax of Green Business Profits

Decision alternative	Maximum payoff	
Half research and half LID (D1)	$30,000	
Research (D2)	($35,000)	← Maximax
Low-impact development (D3)	$34,000	

7.4.5 Determining Minimax Regret

The third part of the problem asks for the minimax regret, which represents the possible opportunity lost. Each profit must be compared to all other profits given that state of nature. The regret is the total amount of money that a decision alternative does not earn, given the maximum profit under that state of nature. For example, the maximum profit with green citizens and government (S1) is $35,000 for D2; given the same state of nature, D3 is expected to result in profit of only $28,000 per month. If the company chose D3 it would miss out on the higher profit from D2; therefore, the regret for choosing D3 is determined to be $7,000 ($35,000 − $28,000). Thus, regret can be considered as an amount equivalent to a lost opportunity. The maximum regret is always considered for each decision alternative under each state of nature. Likewise, the maximum regret for choosing D1 under S2, when one could have earned more with D2, is $5,000. Table 7-5a shows all levels of regret (or opportunity losses) for S1 and S2.

The minimax regret is eventually the decision alternative with the lowest maximum regret. This represents the minimum regret an investor would like to experience if a maximum regret is unavoidable. Hence, the maximum regrets are $5,000 for D1, $14,000 for D2, and $7,000 for D3. Therefore, the minimax regret for this problem is the least of the maximum regrets, which is $5,000 for D1, maintaining the current business strategy (see Table 7-5b). This can also be understood as the lowest amount an investor can expect to lose for making a wrong decision. Hence, should the firm proceed with D1, it will have the least to regret in the event that the decision goes wrong.

Table 7-5a. Regret for Each Decision Alternative under S1 and S2

Decision alternative	Green citizens and government (S1)	Green companies (S2)
Half research and half LID (D1)	$5,000	$4,000
Research (D2)	$0	$14,000
Low-impact development (D3)	$7,000	$0

Table 7-5b. Maximum Regret for Each Decision and Minimax Regret

Decision alternative	Maximum regret	
Half research and half LID (D1)	$5,000	← Minimax regret
Research (D2)	$14,000	
Low-impact development (D3)	$7,000	

7.4.6 Expected Monetary Value

The final part of the problem asks for the *expected monetary value* (EMV) of each decision alternative. The expected monetary value is the sum of the probabilistic values of the payoffs. The EMV can be determined by multiplying the payoff under each state of nature by its probability of occurrence.

The problem stated that each of the two states of nature has an equal chance of occurring, so the probability of each result under each state of nature is equal to 0.5. The EMVs are thus determined by the following equations:

$$\text{EMV D1}: \$30,000 \times (0.5) + \$30,000 \times (0.5) = \$30,000$$

$$\text{EMV D2}: \$35,000 \times (0.5) + \$20,000 \times (0.5) = \$27,500$$

$$\text{EMV D3}: \$28,000 \times (0.5) + \$34,000 \times (0.5) = \mathbf{\$31,000}$$

These calculations show that the largest EMV is $31,000, obtained by choosing decision D3, becoming a low-impact development company. This tells us that the highest expected return would come from going with decision D3.

7.4.7 Expected Opportunity Loss

A variation of the expected monetary value is the *expected opportunity loss* (EOL). Here, the probabilities of occurrence are multiplied by the minimax regret values from Table 7.5a, rather than the values in the payoff table (Table 7.2). The idea of this evaluation is to minimize the expected opportunity loss, rather than maximizing the expected monetary value. Unlike minimax, maximax, and minimax regret, which can result in different recommendations for the optimal decision, the expected opportunity loss calculation will give the same result as the expected monetary value calculation.

Using the given broad probability of 0.5 for each outcome and the values in Table 7.5a, the EOL is calculated as follows:

$$\text{EOL D1}: \$5,000 \times (0.5) + \$4,000 \times (0.5) = \$4,500$$

$$\text{EOL D2}: \$0 \times (0.5) + \$14,000 \times (0.5) = \$7,000$$

$$\text{EOL D3}: \$7,000 \times (0.5) + \$0 \times (0.5) = \mathbf{\$3,500}$$

Because the goal of any decision maker is to minimize the expected opportunity loss, D3 is the preferred choice. As predicted, this decision is the same one as that resulting from the EMV calculations.

7.4.8 Decision Tree Approach

Another risk analysis tool that closely follows the EMV method is a decision tree. This is simply another way of using the EMV method and can often be used to clarify it. Fig. 7-1 shows the decision tree for this problem.

Decision trees can greatly help in solving complex EMV problems and can be a good check on the EMV process, because the answers derived from each method should be the same. As shown in Fig. 7-1, the maximum expected profit still results from D3, with an expected profit of $31,000. This value is derived in the same way as in the equations in section 7.4.6. The two potential profits—$28,000 and $34,000— are each multiplied by their probability of occurrence. These totals are then added together to determine the answer.

This example shows that there are *six* different methods to arrive at an answer to the same problem by using a payoff matrix. The maximin and minimax regret suggest that D1 is the correct answer; the maximax suggests D2; and the EMV, EOL, and decision tree suggest that the owner choose D3. If the owner wants the highest monthly average profit, he should choose decision alternative 3, because it has the highest EMV. However, if the owner prefers to decide based on the best-case scenario, the choice would be D2; if he prefers to decide on the basis of the

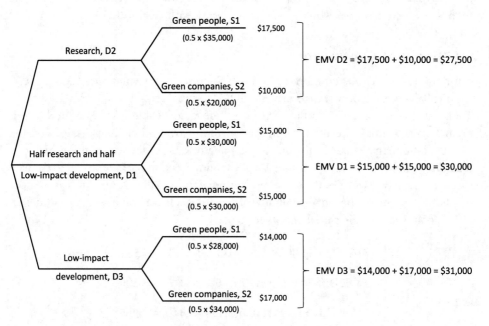

Fig. 7-1. Decision tree to determine EMV

worst-case scenario, the choice would be D1. It all depends on who is the decision maker, and people cannot really be blamed for being who they are. But, for different people and their different perceptions, scientific choices are now available. Without these methods, decision makers would be shooting in the dark and guessing wildly.

Although the payoff matrix offers three objective answers and shows that each decision can be suitable under different decision-making conditions, ultimately the decision maker must make the subjective choice of how to use the information.

7.5 Example Problem 2: Limiting Decision Costs

The second example problem looks at limiting the costs of a decision. The method is basically the same as for Example 1, but the answers will be different because lower outcomes are favored. In this example, an engineering firm is working on a project and wants to keep its data safe from computer viruses. The project is expected to take 30 days to complete, and any computer virus that destroys the data will result in a $50,000 loss. The project manager has three choices for protecting the company's data:

1. Take no action, which will obviously not cost any money.
2. Hire an IT worker at a cost of $12 per hour (minimum) for eight hours per day. This worker will ensure the safety of the data against any potential ordinary viruses.
3. Purchase a safety system for $5,000, which will download all the data to a secure location and ensure its safety against any virus.

The project manager has determined that there is a 70% chance that there will be no obstructive virus during the project, a 25% chance that an ordinary virus could affect the project, and a 5% chance that a super virus could destroy the project. Determine the decision alternatives, states of nature, minimax, minimin, minimax regret, and EMV for the given problem.

7.6 Answer to Example Problem 2

7.6.1 Setting up a Payoff Table

The decision alternatives are the three possible decisions the project manager can make. The first decision alternative (D1) is to do nothing; the second (D2) is to hire an IT worker; and the third (D3) is to purchase a safety system for $5,000. The three states of nature are no virus (S1), ordinary virus (S2), and super virus (S3).

The next step is to determine the cost of each decision alternative given each state of nature. Ample information has been provided; the only remaining information that must be determined is the cost of hiring an IT worker. At $12 per hour for 8 hours a day for 30 days, the cost is $12 \times 8 \times 30 = \$2,880$. Now we set up a payoff table displaying the costs of each decision alternative given each state of nature (Table 7-6).

Table 7-6. Payoff Table: Cost of Keeping Data Safe during 30-Day Engineering Project

Decision alternative	No virus (S1)	Ordinary virus (S2)	Super virus (S3)
Do nothing (D1)	$0	$50,000	$50,000
Hire IT worker (D2)	$2,880	$2,880	$52,880
Safety system (D3)	$5,000	$5,000	$5,000

Any virus will cost the project $50,000 if the data are not properly protected against it. It is believed that the IT worker can prevent any ordinary virus from infecting the system, but this worker will not be able to protect against a super virus. Therefore, the cost to the project if it is hit with a super virus will be the basic damage of $50,000 plus the cost of paying the IT worker, which totals $52,880. The safety system will protect against all viruses, and so its cost under any state of nature will be a constant $5,000.

7.6.2 Determining Minimax

The first part of the question asks for the minimax, which is the best outcome in the worst-case scenario. In this case, *best* means the alternative with the lowest cost. To determine the minimax, start by determining the maximum cost of each decision alternative. The minimum value of all of the corresponding maximum values will be the minimax.

The maximum cost for D1 is $50,000, in case a super virus hits. Similarly, the maximum cost for D2 is $52,880, again if a super virus hits; the worst possible outcome for D3 is $5,000, no matter which state of nature materializes. These values are shown in Table 7-7a. The minimax for this problem is $5,000 for purchasing the safety system; this value is circled in the table. Therefore, in the worst-case scenario, the most optimistic decision will involve the company paying $5,000 to purchase the safety system. In other words, the minimax answers the question of how to make the best of a bad situation.

7.6.3 Determining Minimin

The next part of the question asks for the minimin, which determines what the best choice would be in the best-case scenario so as to mitigate any cost damage that is likely to occur. Different decision strategies can be adopted for different occasions and different conditions. In this problem, the best-case scenario is that there would be no virus, in which case the best choice would have been to do nothing, at a cost of zero dollars. Therefore, the minimin is $0 for the decision alternative of doing nothing and the state of nature of no virus. This is shown in Table 7-7b.

Of course, these are all probabilistic states. We hope that there will be no virus, because that will result in the lowest possible cost, but this is not something we can predict. Calculations are needed to arrive to the answer.

Table 7-7a. Minimax for Cost of Defending against a Virus

Decision alternative	Maximum cost
Do nothing (D1)	$50,000
Hire IT worker (D2)	$52,880
Safety system (D3)	$5,000 ← Minimax

Table 7-7b. Minimin for Cost of Defending against a Virus

Decision alternative	Minimum cost
Do nothing (D1)	$0 ← Minimin
Hire IT worker (D2)	$2,880
Safety system (D3)	$5,000

7.6.4 Determining Minimax Regret

The minimax regret is found in almost the same way as in Example Problem 1. The only difference is that the regret direction is reversed; instead of making the greatest profit, as in Example 1, we now want to pay the lowest cost. The cost of each decision given each state of nature is compared against the best-case scenario for that state of nature. This gives the decision maker the amount of regret for making that decision if that state of nature does occur. Table 7-8a shows the levels of regret.

For S1, the maximum regret is $5,000, which occurs if the safety system is procured and there is no virus; it would have been possible to do nothing, spend nothing, and still face no virus. For S2, the maximum regret that afflicts the project manager is $47,120, which occurs when an ordinary virus hits but the project team has done nothing to prevent it; in contrast, the team could have hired an IT worker for $2,880 who would have saved the computer system from the virus. Lastly, for S3, the maximum regret occurs when, despite hiring an IT worker, a super virus hits the system; to save this expense, the team could have bought the safety system for a mere $5,000 and saved the cost of the damage.

The minimax regret, based on these calculations, is shown in Table 7-8b; it represents the scenario that the project manager will regret least should the worst happen. The minimax regret is the minimum value of the maximum opportunity loss for each decision. Thus, for D1, the maximum opportunity loss is $47,120, if S3 occurs; for D2, the maximum opportunity loss is $47,880, if S3 occurs; and for D3, the maximum opportunity loss is $5,000, if S1 occurs. The minimum of all these maximum opportunity losses (regrets) is $5,000, which is hence the minimax regret for this problem.

Table 7-8a. Regret for Safety of Data against Occurrence of Virus

Decision alternative	No virus (S1)	Ordinary virus (S2)	Super virus (S3)
Do nothing (D1)	$0	$47,120	$45,000
Hire IT worker (D2)	$2,880	$0	$47,880
Safety system (D3)	$5,000	$2,120	$0

Table 7-8b. Minimax Regret for Safety of Data against Occurrence of Virus

Decision alternative	Maximum regret
Do nothing (D1)	$47,120
Hire IT worker (D2)	$47,880
Safety system (D3)	$5,000 ← Minimax regret

7.6.5 EMV

More than minimin, minimax, or minimax regret—which can all be valid decision-making strategies, depending on the risk-taking profile of the decision maker—EMV may very well be more appropriate because it takes into account the probabilities of events.

The project manager has determined that the probability that the data will be hit by no virus is 0.7; by an ordinary virus, 0.25; and by a super virus, 0.05. The EMV of each decision alternative is determined by multiplying each cost for a state of nature by its probability. All of these costs are then summed to determine the EMV of the decision alternative. The EMVs are thus calculated as follows:

$$\text{EMV D1}: \$0 \times (0.7) + \$50,000 \times (0.25) + \$50,000 \times (0.05) = \$15,000$$

$$\text{EMV D2}: \$2,280 \times (0.7) + \$2,280 \times (0.25) + \$52,880 \times (0.05) = \$4,810$$

$$\text{EMV D3}: \$5,000 \times (0.7) + \$5,000 \times (0.25) + \$5,000 \times (0.05) = \$5,000$$

Therefore, the lowest expected cost, given the probabilities of all conditions, is to hire the IT worker for an EMV of $4,810.

It is now up to the project manager which option to choose. Does this manager wish to take as little risk as possible and choose based on the minimax regret? Is the manager just going to cross his fingers, hope for the best, and choose based on the minimin? Or is the manager going to use the probabilities to determine the best choice based on EMV? It depends on the decision maker's tolerance for risk, but the payoff matrix technique displays all of the different types of risks and gives them numerical values.

7.7 Example Problem 3: Qualitative Decision Making

The third and final payoff matrix example deals with qualitative instead of quantitative decision making. A managerial position recently opened up in an engineering firm, and the firm's owner needs to decide among three lower-level engineers for the promotion. The three candidates have different strengths and weaknesses, and the owner must decide which characteristics will be best for making a good engineering manager. She has decided to weigh the candidates' strengths in three categories: The first category represents the candidates' skills as an engineer, the second category weighs each candidate's leadership qualities, and the final category considers the candidates' ability to work with others.[2]

The owner has decided that not all talents are equal in determining an effective manager. She wants the manager to be a well-versed engineer, so she has given engineering skills a weight of 0.5. The owner views leadership as slightly less important and has assigned it a weight of 0.3. Finally, she has assigned the candidates' ability to work with others a weight of 0.2. Each candidate has been graded in these three areas, and Table 7-9 shows the resulting payoff matrix.

Whom should the owner select? Use payoff matrix techniques to discover a suitable answer.

7.8 Answer to Example Problem 3

To use payoff matrices, the maximin, maximax, minimax regret, and EMV must be evaluated for the given problem. Naturally, the ability qualities must be maximized so that the owner can find the best candidate.

7.8.1 Maximin

Start by determining the minimum quality for each candidate and the corresponding numerical grade. This calculation will find the maximum of the minimum, so that the owner can make the best of a bad situation. The minimum quality for candidate 1 is a teamwork grade of 1 ("a real jerk"). Nobody wants to work with a jerk, and a jerk is not someone ideal for working directly with clients. The minimum quality for candidate 2 is engineering skills, with a grade of 5, and this doesn't sound technically sufficient for an engineering job. The minimum quality for candidate 3 is leadership, with a grade of 7. Hence, it is clear that the maximum of all these minimum values is 7, which is thus the maximin. By choosing candidate 3, the owner would be able to select the best of the worst. By doing this, she is able to cut her losses in case the worst happens. Thus, she can avoid the worst outcome if the worst conditions were to materialize.

---

[2]Despite all the technical learning in engineering school, it is actually a fact that most of the human skills needed in an engineering job are simply not taught in schools.

Table 7-9. Candidates' Grades in Engineering Skills, Leadership, and Teamwork

Candidate	Engineering skills (S1)	Leadership (S2)	Teamwork (S3)
Candidate 1 (D1)	10	3	1
Candidate 2 (D2)	5	9	9
Candidate 3 (D3)	8	7	8

Note: The decision alternatives in this problem are the three different candidates that the owner is considering for the promotion. The states of nature that can kick into action are engineering skills, leadership, and pleasantness. The grades are on a scale of 1–10 with 10 being the highest.

7.8.2 Maximax

To see who might be the best of the best, find the maximax. In the event that the owner is able to get the best out of all candidates, she can go with the candidate who excels most in any characteristic. The owner can aim to work with the strengths of the candidates, rather than mitigate their weaknesses.

But which candidate stands out the most in any given category? Looking at the chart, it becomes clear that the maximax is 10, because this is the highest grade for any candidate in any state of nature. This grade is for candidate 1 in the category of engineering skills. If the owner is looking for a candidate who excels in any one category, she would use the maximax and choose candidate 1. With some good luck, the engineering and technical design part of her business could thrive.

7.8.3 Minimax Regret

But now the owner can look at the minimax regret. This will serve to minimize the maximum regret the owner would feel if the situation were to deteriorate. When dealing with a question like this, the minimax regret indicates which decision would leave the decision maker with the person who has no glaring weaknesses compared to the others.

For S1–D1, there is no regret. For S1–D2, the owner could experience a regret score of 5 were she to promote candidate 2, when she could have promoted candidate 1, with higher technical capabilities. Likewise, for S1–D3, the owner's regret is 2 points because she could have promoted candidate 1 instead of candidate 3.

This result could also be interpreted in another way by the owner. She sees that in engineering skills, the maximum grade is 10 (candidate 1), whereas candidate 3 has a grade of 8, meaning the owner will have a regret level of 2. In this manner, we can go down the S2 column and S3 columns, arriving at Table 7-10a, showing the level of regret for each candidate based on each state of nature. For instance, the candidate with the strongest leadership qualities is candidate 2, with a grade of 9; candidate 3 has 7 leadership points, so he trails by only 2 points. As far as leadership regret is concerned, the largest regret would be by choosing candidate 1, followed by candidate 3. Likewise, the maximum regret for teamwork would be for candidate 1 again, followed at a distance by candidate 3.

Table 7-10a. Regret Levels (Opportunity Loss) for the Candidates based on Engineering Skills, Leadership, and Teamwork

Candidate	Engineering skills	Leadership	Teamwork
1	0	6	8
2	5	0	0
3	2	2	1

Table 7-10b. Maximum Regret and Minimax Regret for Candidate Selection

Decision alternative	Maximum regret	
Candidate 1	8	
Candidate 2	5	Minimax regret
Candidate 3	2	

That said, the maximum regret with taking a decision by going with candidate 1 is 8 points, on the teamwork score; the maximum regret for candidate 2 is 5 for the engineering skills score; and the maximum regret for candidate 3 is 2, with both engineering skills and leadership skills. The maximum regret table for each candidate is presented in Table 7-10b. From here, it is seen that the minimum of all the maximum regrets is 2, belonging to candidate 3. Hence, the safest bet for regretting the least in the end is by choosing candidate 3. The minimax regret is circled in Table 7-10b.

7.8.4 EMV

In this case, the EMV is the expected ability of each candidate. The owner has assigned a probability (in this case, a weight) for engineering skills of 0.5; leadership has been assigned a weight of 0.3; and teamwork is given a weight of 0.2, although that might be the most vital talent in getting along well with people both internal and external to the company. The expected ability of each candidate can then be determined as follows:

$$\text{EMV D1}: 10 \times (0.5) + 3 \times (0.3) + 1 \times (0.2) = 6.1$$

$$\text{EMV D2}: 5 \times (0.5) + 9 \times (0.3) + 9 \times (0.2) = 7.0$$

$$\text{EMV D3}: 8 \times (0.5) + 7 \times (0.3) + 8 \times (0.2) = 7.7$$

Based on the EMVs, the owner should promote candidate 3, because he brings the maximum of the probabilistic scores (EMV = 7.7) to the company.

Again, the owner may choose to promote a candidate based on any of the other three characteristics. Candidate 1 is clearly the best engineer, and on that basis alone would appear to be the best choice. However, a perceptive owner should realize that it is just as important that a manager have effective people skills. A manager needs to serve as a strong leader for employees but also show compassion when necessary and have enough people skills to work well with clients.

A payoff matrix allows the employer to give each skill a weight and objectively choose whom to promote.

7.9 Importance of the Payoff Matrix

The final result of a payoff matrix gives the decision maker four to six possible answers, depending on all the exercises undertaken. The decision maker can then choose based on what type of risk she is willing to take. If the decision maker has an optimistic view and wants to choose based on the best-case scenario she will use the maximax. If she is pessimistic and wants to plan for the worst she will decide based on the maximin, which allows her to look at the worst-case scenario for each decision alternative and choose the best one. The individual who chooses based on the maximin would like to walk away with something positive and productive, no matter what the state of nature is. This may not have the potential offered by the maximax, but it is unlikely to result in great disappointment either. A similar decision-making method, the minimax regret, is best for the individual who does not want to miss out too much in any one category. It is a well-balanced decision that, once again, may not have the highest potential gain but does have the lowest opportunity for loss.

The final method of using the payoff matrix is perhaps the most important. Determining the expected monetary value (EMV) based on probabilities allows all states of nature to be included in the final decision, in contrast to previous methods where decisions are made by excluding some states of nature. With EMV, the decision maker knows the probability of each outcome and can determine, overall, what he or she can expect to receive from that decision.

People use processes similar to payoff matrices in nearly every decision that they make. Which method they choose depends on the decision to be made and the type of people they are. Small decisions are often made based solely on the maximax of each decision alternative. For example, when choosing a snack out of the vending machine, people often choose what they think is the most delicious. When doing so, they are unconsciously making a decision based on the maximax. If people factor into their decision states of nature such as health benefits, taste, and cost—perhaps even unconsciously assigning weights to each of their choices—then their decision more closely resembles that of an EMV problem. Maybe somebody thinks that the cookies will taste the best but knows that they contain a lot of fat and cost 50 cents more than the crackers; the crackers may not taste as good, but are less expensive and healthier. The decision then would come down to the individual and how much weight he or she has assigned to each category.

When a person faces a larger or life-changing situation, the payoff matrix method becomes even more important. A huge decision, such as where to live, involves many decision alternatives and nearly as many states of nature. Some people will make this decision based on the maximax. For example, a man who now lives in Hawaii perhaps one day said before moving to Hawaii, "I want to live somewhere with great weather." That was the only criterion that really mattered, and so his decision was based on the maximax. Maybe a woman also wants to live in good weather but does not want to live too far from her family, so she compromises and moves to Florida. This individual does not want a lot of regret, and so she made the decision based on the minimax regret method. Other people might even be serious enough to assign each state of nature a probability that it will lead to increased happiness. They may weight weather, closeness to family, job market, and other factors and use the EMV method to make their ultimate decision.

In general, when making important decisions in risk management in engineering, the EMV method is recommended whenever possible. Risk-related decisions in engineering can involve multimillion-dollar decisions. The potential outcomes should be known so that decision makers can assign probabilities to each occurrence. By knowing the expected cost of each decision alternative, companies are able to make more informed decisions. Making decisions based on known probabilities will ultimately lead to a much more successful business.

Overall, the payoff matrix method is great for decision makers with any attitude toward risk. Risk takers, risk avoiders, and those in between can all use this technique to come up with answers that suit their needs. Each method can also be combined with one of the other methods to help reassure the decision maker. If a decision yields the same outcome for maximin, maximax, minimax regret, and EMV, the choice will be obvious. However, most decisions are not so cut and dried, so the decision maker must choose what type of risk to incur.

7.10 Exercises

Problem 1

If the following payoff matrix represents profits, what are the (a) maximin, (b) maximax, and (c) minimax regret?

	S1	S2
D1	5	8
D2	6	12
D3	7	9

Problem 2

If the following payoff matrix represents costs, choose the best decision based on (a) minimax, (b) minimin, and (c) minimax regret.

	S1	S2
D1	2	4
D2	5	6
D3	4	7

Problem 3a

Determine the EMV for Problem 1, given that the probability of S1 = 0.6 and the probability of S2 = 0.4.

Problem 3b

Draw a decision tree for Problem 1 using the probabilities given in part (a) of this problem.

Problem 4

When making an important decision in your own life, which payoff matrix method (maximin, maximax, minimax regret, or EMV) do you prefer to use? Why?

Problem 5a

Determine the decision alternatives and states of nature for the following situation: A family has decided to buy one of three types of puppy: golden retriever, Labrador retriever, or pit bull. The family wants a dog that will serve as a good companion and playmate for the children but would also like a dog that could protect the house if burglars broke in.

Problem 5b

Given these decision alternatives and states of nature, which dog would you choose? Construct a payoff matrix table with your own values to back up your decision.

CHAPTER 8

Bayes' Theorem

8.1 Introduction

Often it is difficult to discern the likely cause of an outcome, but doing so is essential to diagnosing the problem so that remedial measures and intervention may be introduced. One needs to resort to techniques, such as Bayes' theorem, that are able to mathematically derive the likelihood of events that could cause the specific outcome. In Chapter 4, we studied cause-effect diagrams where causes had very discrete, nonprobabilistic connections to outcomes. Here, although we know that a certain event might be the cause of an outcome, we need to determine the likelihood of its being the real cause.

The purpose of this chapter is to go over the derivation of Bayes' theorem from conditional probability, followed by example applications to demonstrate our understanding of the theorem. Sample exercises demonstrate applications of Bayes' theorem that should help students to better understand the theorem through practice.

Bayes' theorem was derived from conditional probability through the understanding of events and sample space, the multiplication rule, and the law of total probability. The example applications and exercises show that known beliefs can be updated with the use of current data. Furthermore, the applications through the examples prove that the theorem is an excellent tool for risk analysis in the decision-making process.

8.2 What is Bayes' Theorem?

Bayes' theorem is specifically derived from conditional probability, which measures the probability that an event will occur, given that another, separate event has occurred beforehand. According to Vanem (2013), the theorem "expresses how a subjective degree of belief should rationally change in light of evidence or data" (p. 14). In other words, Bayes' theorem is a mathematical expression of learning from past experiences.

To formulate and determine the conditional probabilities, Thomas Bayes[1] introduced a priori and a posteriori probability fundamentals. The a priori represents the *known*, prior probability distributions; the a posteriori is the new or revised probability distribution, given the conditional—or subjected—circumstances.

Prior probabilities are estimates of an event occurring before any decision is made. These estimates are based on subjective evaluations, prior beliefs, collateral information, and intuition. Therefore, it requires an individual with a great depth of knowledge and experience to determine probabilities and make decisions (Freund and Williams 1977). Having derived the prior probability from knowledge or degree of belief, Bayes' theorem allows computation of the probability of the outcome when only incomplete information (i.e., uncertainties) is available, which is the case in most decision-making problems (Ahmed et al. 2005, Das 1999).

Bayes' theorem can be applied to a variety of industries and fields, such as construction, medicine, industry, organizational management, and international relations. Since Bayes' theorem primarily quantifies the probability that an event will occur, it can be a useful tool for risk analysis, estimation, and any decision-making process.

8.3 Criticism and Advantages

Throughout the years, the validity of Bayes' theorem has been questioned because it is based on subjective interpretation and belief. Bayesian inferences require knowledge and skill to translate these prior beliefs into something more scientific and less subjective. There are no standard procedures to explain how to determine a prior probability, and there is no correct way to choose one. If prior probabilities are not selected by an experienced individual, he or she can generate misleading results. Thus, Bayes' theorem can produce posterior probability distributions that are biased by the prior probabilities (SAS Institute 2010).

That said, Bayes' theorem is mathematically perfect and constituted an advance in mathematical science when first presented. In addition, its mathematical simplicity is considerable. If the prior probabilities can be applied accurately, Bayes' theorem yields very high quality, reliable information.

8.4 Derivation of Bayes' Theorem from First Principles

8.4.1 Events and Sample Space

To derive Bayes' theorem, one must first understand the concepts of *events, sample space,* and *conditional probability* and possess a basic knowledge of statistics. This section discusses events and sample space.

---

[1]Thomas Bayes, 1701–1761. Bayes was an English philosopher, Presbyterian minister, and statistician (Belhouse 2016). He is known for having published *Divine Benevolence, or an Attempt to Prove That the Principal End of the Divine Providence and Government is the Happiness of His Creatures (1731).* Bayes' theorem was never published by him and only became known after his death.

The customary symbols are read as follows: ∪ as "union" or "or"; ∩ as "intersection" or "and"; A' as "not A"; and "|" as "given."

A set of points representing all possible outcomes is known as a sample space. An event is an individual outcome or set of outcomes in a given, finite sample space. Probability is the likelihood that a certain event will occur. When dealing with probabilities and how they should behave in a finite sample space, one must define three basic postulates from the laws of probability (Freund and Williams 1977):

1) $0 \leq P(A) \leq 1$, where the probability of event A is between 0 and 1.
2) $P(S) = 1$, where the sum of the probabilities of all outcomes in the sample space, S, must equal to 1.
3) Two events A and B are mutually exclusive if

$$P(A \cup B) = P(A) + P(B) \quad \text{and} \quad P(A \cap B) = 0.$$

Figs. 8-1 to 8-3 will provide a better understanding of the relationship between events, sample space, and probability.

Fig. 8-1 represents the finite sample space S as a rectangular box encompassing the circle representing event A. Fig. 8-2 shows the nonoccurrence of event A, written as A', in the hatched area within the sample space. Note that the circle representing event A is no longer shaded. If $P(S) = 1$, then $A' = 1 - A$. Alternately, we can say that $A + A' = 1$.

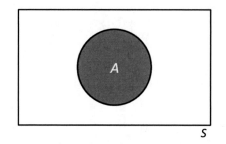

Fig. 8-1. Event A in sample space S

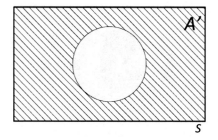

Fig. 8-2. Nonoccurrence of event A (or occurrence of A')

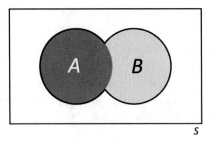

Fig. 8-3. *A* ∪ *B* within the sample space

Multiple events that occur can interact with each other in a given sample space. These interactions are best displayed in Venn diagrams. In the following cases, two events, *A* and *B,* interact within the sample space. The sum, or union, of these two events is denoted as *A*∪*B*, which represents the combined probability of the events—that is, the probability that event *A* or event *B* or both will occur—and is illustrated in Fig. 8-3; the events are outlined to represent their union. The probability of the union of *A* and *B* in a sample space is denoted as *P*(*A*∪*B*). Because this is a function of the sample space, *P*(*A*∪*B*) implies that it is the probability of *A*∪*B* given the sample space *S*, or *P*(*A*∪*B*)|*S*.

Fig. 8-4 illustrates the outcomes of what the two events have in common. This is represented by the central crosshatched area where events *A* and *B* intersect and is inherently known as the intersection, or product, of the events. This is denoted as *A*∩*B* and its probability as *P*(*A*∩*B*). Or, similar to the previous discussion for union, the intersection of events *A* and *B* given the sample space *S* can be written *P*(*A*∩*B*)|*S*. *P*(*A*∩*B*) can also be referred to as the joint probability of events *A* and *B*.

8.4.2 Conditional Probability and Dependence

Given these event relationships within the sample space, we can now establish the conditional probabilities. If events are not independent, then they are considered to be dependent. If events are independent, the occurrence of one has no effect on the

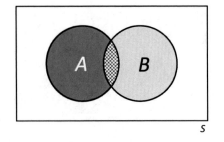

Fig. 8-4. *A* ∩ *B* within the sample space

probability of the other; if they are dependent, the probability of the occurrence of one event is affected by the occurrence or nonoccurrence of the other. The following equations, derived from the laws of probability, are relevant to conditional probability and the derivation of Bayes' theorem (Freund and Williams 1977).

The formula for independent events is derived from the *special multiplication rule*

$$P(A \cap B) = P(A) \cdot P(B)$$

The formula for dependent events is derived from the *general multiplication rule*

$$P(A \cap B) = P(B) \cdot P(A|B)$$

Because it does not matter which event is referred to as A or B, we can say that

$$P(A \cap B) = P(B) \cdot P(A|B) = P(A) \cdot P(B|A)$$

which is Bayes' theorem that deals specifically with dependent events and allows for the updating of probability values should new information arise. In the previous equation for the *general multiplication rule*, we see the conditional probability $P(A|B)$. This is defined as the probability that event A will occur given that event B has occurred. This conditional probability is where the new information would be applied to revise the posterior probability.

8.4.3 Bayes' Theorem Derived from Conditional Probability

To illustrate how Bayes' theorem can be derived simply from conditional probability, let us consider the following example:

> In a given year, 80 civil engineering students at a local university apply for a summer internship at a prestigious engineering firm. Of all the applicants, 35 have prior internship experience, whereas the rest do not. Of the 50 applicants considered to be upper-class students (juniors and seniors), 20 have prior internship experience. There is only one position available. What is the probability that an upper-class student, given that they had no prior experience, will be hired for the open position?

The data given can be tabulated as shown in Table 8-1, where U denotes the selection of an upper classman, L denotes the selection of a lower classman,

Table 8-1. Initial Table for Conditional Probability Example

	Experienced (E)	No experience (N)	Total
Upper Classmen (U)	20	—	50
Lower Classmen (L)	—	—	—
Total	35	—	80

E denotes the selection of a student with prior experience, and N denotes the selection of a student with no experience. Using the given data, we can fill the remaining cells with a bit of arithmetic, as shown in Table 8-2.

The individual probability of selecting a specific student applicant can now be determined by making the following calculations. The probability of selecting an upper classman from the entire sample is

$$P(U) = \frac{20 + 30}{80} = 0.625$$

The probability of selecting a lower classman is

$$P(L) = \frac{15 + 15}{80} = 0.375$$

The probability of selecting a student with prior internship experience is

$$P(E) = \frac{20 + 15}{80} = 0.4375$$

The probability of selecting a student with no prior internship experience is

$$P(N) = \frac{30 + 15}{80} = 0.5625$$

The probability of selecting an upper classman and a student with no experience is

$$P(U \cap N) = \frac{30}{80} = 0.375$$

Therefore, to determine the probability that an upper classman is selected, if given that he or she has no prior experience and assuming that each applicant has an equal chance (1/80) of being hired, we compute

$$P(U|N) = \frac{30}{45} = 0.667 = 66.7\%$$

Table 8-2. Final Table for Conditional Probability Example

	Experienced (E)	No experience (N)	Total
Upper Classmen (U)	20	30	50
Lower Classmen (L)	15	15	30
Total	35	45	80

Note that this conditional probability can be also written as

$$P(U|N) = \frac{30/80}{45/80} = \frac{P(U \cap N)}{P(N)} = \frac{30}{45} = 0.667$$

Therefore,

$$P(U \cap N) = P(U|N) \cdot P(N)$$

Similarly, $P(N|U) \cdot P(U) = P(N \cap U)$. Because $P(U \cap N) = P(N \cap U)$, we can equate the two formulas. Thus,

$$P(U|N \cdot (P(N)) = P(N|U) \cdot P(U)$$

And, in general,

$$P(A|B) \cdot P(B) = P(B|A) \cdot P(A)$$

which is Bayes' theorem.

8.5 Logic of Prior/Posterior Probabilities and Tree Diagrams

The logic of Bayes' theorem can be applied in either direction. The first logic to consider would be from *cause to effect*, as shown in Fig. 8-5. Event A represents the cause, and event B represents the effect, or outcome. The probabilities that follow this logic are the a priori, or prior, probability and its respective conditional probabilities. The a priori probabilities are used to derive the a posteriori probabilities. Conversely, reasoning in the opposite direction would go from *effect to cause*, using the a posteriori, or posterior, probability and its conditional probabilities as a check to confirm the a priori probability values. This *effect to cause* logic, applied to the example in Fig. 8-5, can now be seen in Fig. 8-6, where events B and B' are now the probable causes, and events $A, A_1 \ldots A_k$ are now the probable effects. Generally, when going from effect to cause, the a posteriori becomes the new a priori. Whether one starts with a priori or a posteriori is often a function of the type of data available, but they are both mutually interrelated.

Tree diagrams (Figs. 8-5 and 8-6) are used in conjunction with Bayes' theorem for multiple, mutually exclusive events. When events are *mutually exclusive*, it means that the events cannot occur at the same time. As shown in Fig. 8-5, events $A, A_1 \ldots A_k$ are the mutually exclusive causes, where A is one event, A_1 is the second event, and A_k represents a subsequent event, and events B and B' are the mutually exclusive effects/outcomes. Conversely, in Fig. 8-6, events B and B' are the mutually exclusive causes, and events $A, A_1 \ldots A_k$ are the mutually exclusive effects. The prime shown in event B' indicates its distinction and mutual exclusivity from B.

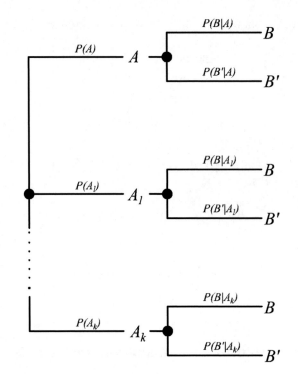

Fig. 8-5. General/prior (a priori) tree diagram

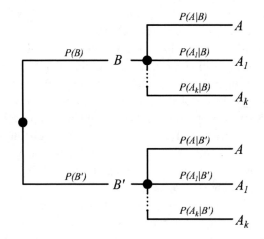

Fig. 8-6. Posterior (a posteriori) tree diagram

These trees are a useful aid when visualizing the logic behind the processes, especially when considering more than two causes and effects. Hence, this chapter builds on earlier chapters that used trees (Ch. 6) and cause-effect diagrams (Ch. 4). All together, these three chapters complement each other.

There are three characteristics to consider about these tree diagrams:

1) Branches stem from a central node (shown as a dot on the tree diagrams).
2) The total probability of all of the branches emanating from a node must be equal to 1.0, or 100%.
3) The events assigned to branches stemming from a node must be mutually exclusive.

The general formula for Bayes' theorem, when considering multiple, mutually exclusive events, is an expansion of what was previously derived:

$$P(A_i|B) = \frac{P(A_i) \cdot P(B|A_i)}{P(A_1) \cdot P(B|A_1) + P(A_2) \cdot P(B|A_2) + \cdots + P(A_k) \cdot P(B|A_k)}$$

for $i = 1, 2, \cdots, k$.

where $P(A_1) \cdot P(B|A_1)$ is equal to the joint probability of reaching B from A_1; $P(A_2) \cdot P(B|A_2)$ is equal to the joint probability of reaching B from A_2; and $P(A_k) \cdot P(B|A_k)$ is equal to the joint probability of reaching B from A_k. The entire denominator is equal to the sum of all the joint probabilities, which is $P(B)$ in the above formula.

Tabulating these joint probabilities further aid in their calculation. The use and application of these tree diagrams and prior and posterior probability calculation tables are shown in Sections 8.6.1 and 8.6.2.

A caveat to consider when calculating the prior and posterior probabilities in the tables is the decimal rounding of these probabilities. Probabilities may or may not result in the exact calculated values, due to errors in rounding. An example of this situation also is shown in Sections 8.6.1 and 8.6.2. Keep in mind that the total probability of events in a given sample space must equal 1, or 100%.

8.6 Examples and Sample Exercises: Applications of Bayes' Theorem

Now that Bayes' theorem has been derived, the following sections cover the application of Bayes' theorem through a series of simple examples of real-world situations. These sections also cover the setup and use of the a priori and a posteriori probability calculation tables.

In the first example, a client firm seeks probabilistic information on a particular contractor for an upcoming project. The contractor has a great reputation and does quality work, but has projects that run over budget and are delayed. The client firm wants detailed information on this contractor to be able to compare with other contractors. In this scenario, Bayes' theorem can help in the decision-making process by enabling the quantification and comparison of performance attributes, thereby helping to reduce risk.

The second example deals with production, management, and industrial applications. Again, Bayes' theorem can prove useful in risk assessment and decision making. In this scenario, a firearms manufacturer outsources several ancillary parts for its rifle. Although firearms are built within tight tolerances, no system is truly perfect; therefore, the parts are susceptible to manufacturing defects. Knowing the chances of defective parts occurring and being able to quantify the probability of a particular part having an increased risk of defects would be beneficial to the manufacturer. In other words, having an idea of what parts could be defective can help the manufacturer decide which parts need to have an increase in inventory, thereby reducing the risk of costs caused by delayed production because of having an insufficient quantity of a particular part on hand for a complete rifle.

8.6.1 Example 1: Contractor Performance Information

A client firm needs to decide on whether they should consider to hire a particular contractor for an upcoming project. For this, they need probabilistic information on pssible delays and budget performance. The contractor, All-Win Construction, has a good reputation, produces quality work, and typically complete their projects, within budget, at a rate of 95%. Past information reveals that the probability at which All-Win Construction encounters a delay, given that they complete a project within budget is 10%. On the other hand the probability that All-Win Construction encountered a delay, given that they went over budget is 70%.

Now, if All-Win Construction were to encounter a delay, what is the probability that they will complete a project within budget?[2] Help the client firm arrive at this information.

The first step, therefore, is to assign variables to the given events:

- C = Completed projects within budget
- N = Completed projects over budget (not within budget)
- D = Encountered delays
- ND = Encountered no delays

From here, a tree diagram can be created to organize the events given in the problem (Fig. 8-7). Each branch represents an event, has its own probability, and is typically read from left to right.

Take branch C in Fig. 8-7 as an example. The probability, or a priori, of branch C is $P(C)$. Branches D and ND stem from branch C. Subsequently, going from branch C to branch D yields a conditional probability, $P(D|C)$; correspondingly, going from branch C to branch ND yields $P(ND|C)$.

---

[2]This can be rephrased: What is the probability that the contractor will complete within budget, given that a delay was encountered?

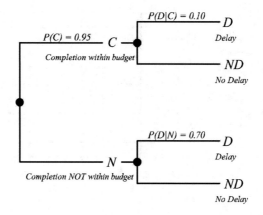

Fig. 8-7. Initial tree diagram with given events for Example 1

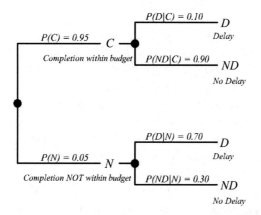

Fig. 8-8. Final (a priori) tree diagram for Example 1

Fig. 8-8 shows the tree diagram with the given probabilities, along with the complementary probability of each corresponding branch. The total probabilities of corresponding branches must equal 1, or 100%. For example, in Fig. 8-8, branch C splits off into branches D and ND. The conditional probability for branch ND is $P(ND|C) = 0.90$, because the conditional probability given for branch D is $P(D|C) = 0.1$. It is thus seen that $P(D|C) + P(ND|C) = 1 \Rightarrow P(ND|C) = 1 - P(D|C) = 0.90$. The same applies to the branches stemming from branch N.

The individual probabilities of branches D and ND can now be calculated.

Fig. 8-9 shows the calculation of these probabilities; the last row of the table is simply the sum of the probabilities for each branch. Keep in mind that the sum of these two probabilities must equal 1. This figure extends Fig. 8-8 to simulate the probabilities of D and ND. Coming from $P(D|C)$, there can be no ND possible; coming from $P(ND|C)$, there can be no D possible, and so forth. This is reflected in Fig. 8-9.

Now, Bayes' theorem can be applied to determine the a posteriori probability that the contractor will complete the project within budget, given that they encountered a delay, or $P(C|D)$.

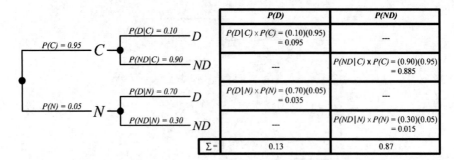

	P(D)	P(ND)
$P(D|C) \times P(C) = (0.10)(0.95)$ $= 0.095$		
	---	$P(ND|C) \times P(C) = (0.90)(0.95)$ $= 0.885$
$P(D|N) \times P(N) = (0.70)(0.05)$ $= 0.035$		---
	---	$P(ND|N) \times P(N) = (0.30)(0.05)$ $= 0.015$
$\Sigma =$	0.13	0.87

Fig. 8-9. Example 1 probability calculations for a priori $P(D)$ and $P(ND)$

$$P(C|D) = \frac{P(D|C) \cdot (C)}{P(D)} = \frac{0.095}{0.13} = 0.73 = \mathbf{73\%}$$

Thus, the probability that All-Win Construction will complete the project within budget, given that a delay is encountered is 73%. At this point, the client firm might take into consideration other attributes of All-Win Construction or consider another contractor because 73% is not a stellar rating.

To check these calculations, let us calculate the posterior probabilities (a posteriori) by building the tree diagram in reverse (or inside-out), as shown in Fig. 8-10. The tree is set up similarly to how it was set up previously, except for using the newly calculated probabilities for D and ND. The conditional probabilities, which were initially the a priori probabilities, can now be back-calculated using Bayes' Theorem using the data from the a priori calculations. If the calculations are all correct, the posterior probabilities of C and D should equal the prior probabilities of C and D with which we originally started. The calculations for these conditional posterior probabilities are as follows:

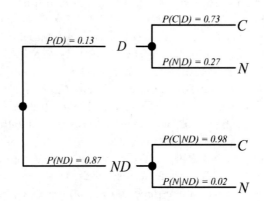

Fig. 8-10. A posteriori tree diagram for Example 1

$$P(C|D) = \frac{P(D|C) \cdot P(C)}{P(D)} = \frac{0.095}{0.13} = 0.73$$

$$P(N|D) = \frac{P(D|N) \cdot P(N)}{P(D)} = \frac{0.035}{0.13} = 0.27$$

$$P(C|ND) = \frac{P(ND|C) \cdot P(C)}{P(ND)} = \frac{0.855}{0.87} = 0.98$$

$$P(N|ND) = \frac{P(ND|N) \cdot P(N)}{P(ND)} = \frac{0.015}{0.87} = 0.02$$

Labeling the tree diagram with the appropriate values and probabilities yields the tree shown in Fig. 8-10. Once again, we see that $P(C|D) + P(N|D) = 1.0$, as is $P(C|ND) + P(N|ND)$.

As in the a priori calculations, the individual probabilities of branches C and N can now be calculated. The branches are extended in Fig. 8-11 with an aim to discover these probabilities of C and N. Observe, again, that $P(C) + P(N) = 0.95 + 0.05 = 1$. This means that the calculations are accurate. If the values are off by a considerable margin, then it is advisable to recheck the calculations; ignore or adjust minor rounding errors.

8.6.2 Example 2: Defective Parts at a Firearms Manufacturer

A firearms manufacturer orders different rifle parts from various suppliers, because it does not produce these specific parts in-house. These parts are built within strict tolerances, and defects are rare. Unfortunately, some defects do slip through quality control and end up being shipped out in the final product.

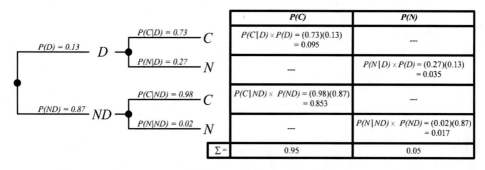

Fig. 8-11. A posteriori probability calculations for Example 1

Supplier A makes the trigger assemblies of which, records show, about 1% have been defective. Supplier B makes the rifle butt stocks, of which 3% have been defective; Supplier C makes the slings, of which 2% have been defective; and Supplier D makes the forward grips, of which 5% have been defective. The rifle manufacturer's current inventory is shown in Table 8-3.

Choosing one part at random, what is the probability that it is defective? Also, which part would have the greatest likelihood of being defective?

First, the a priori must be determined before the tree diagram can be created. To do this, start by taking the total number of units for each part and dividing that by the total number of parts (2,550 units). For example, the manufacturer has 400 units of Supplier A's trigger assemblies. Therefore, the probability of a trigger assembly being chosen at random is $400/2,550 = 15.7\%$.

The remaining probabilities are calculated similarly and shown in Table 8-4. The conditional probabilities can be deduced from the problem statement. For example, let $P(F|A)$ represent the probability that the part is defective, given that it is from Supplier A.

Again, before a tree diagram is created, letters must first be assigned to represent the corresponding events:

- A = Supplier A; trigger assemblies
- B = Supplier B; butt stocks
- C = Supplier C; slings
- D = Supplier D; forward grips
- F = Defective
- ND = Not Defective

Using the data that was given and the data that was calculated, the tree diagram can now be created (Fig. 8-12). Applying the probability calculations used in the

Table 8-3. Current Inventory Held by Manufacturer

Trigger assemblies (A)	Rifle butt stocks (B)	Slings (C)	Forward grips (D)
400	500	1,000	650
Total number of parts = $400 + 500 + 1,000 + 650 = 2,550$			

Table 8-4. a Priori for all Suppliers

Supplier A	Supplier B	Supplier C	Supplier D
15.70%	19.60%	39.20%	25.50%

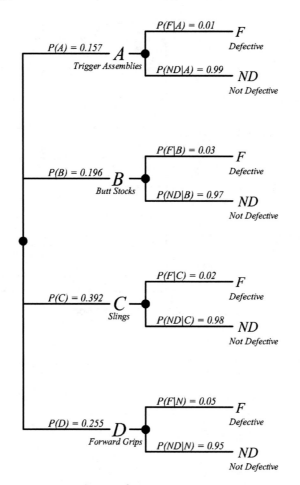

Fig. 8-12. A priori tree diagram for Example 2

previous example, the probability of choosing a defective part, $P(F)$, is 2.8%. The calculations can be seen in Fig. 8-13.

To determine the part that would most likely be defective if randomly selected, apply Bayes' theorem to the example to find a posteriori probabilities:

$$P(A|F) = \frac{P(F|A) \cdot P(A)}{P(F)}$$

Solving for Supplier A:

$$P(A|F) = \frac{P(F|A) \cdot P(A)}{P(F)} = \frac{(0.01)(0.157)}{0.028} = 0.056 = 5.6\%$$

And likewise for the remaining suppliers:

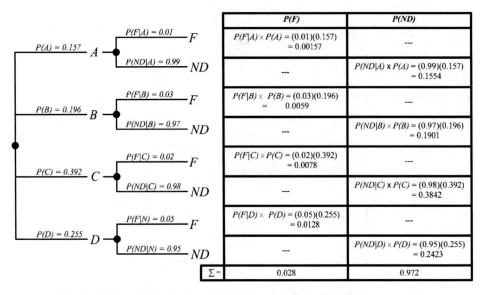

Fig. 8-13. Extended tree for calculating outcome probabilities

$$P(B|F) = \frac{P(F|B) \cdot P(B)}{P(D')} = \frac{(0.03)(0.196)}{0.028} = 0.210 = 21.0\%$$

$$P(C|F) = \frac{P(F|C) \cdot P(C)}{P(F)} = \frac{(0.02)(0.392)}{0.028} = 0.280 = 28.0\%$$

$$P(D|F) = \frac{P(F|D) \cdot P(D)}{P(F)} = \frac{(0.05)(0.255)}{0.028} = 0.454 = \textbf{45.4\%}$$

Check that $P(A|F) + P(B|F) + P(C|F) + P(D|F) = 1$:

$$P(A|F) + P(B|F) + P(C|F) + P(D|F) = 0.056 + 0.210 + 0.280 + 0.454 = 1 \checkmark$$

Based on these calculations using Bayes' theorem, the part that will most likely be defective when selecting at random is the forward grip made by Supplier D, with a probability of 45.4%. In words, the probability that defects are in the trigger assemblies given that defects are found is 5.6%; that defects are in the butt stocks given that defects are found is 21%; that defects are in the slings given that defects are found is 28%; and that defects are in the forward grips given that defects are found is 45.4%. Applying the same methods from the previous example, the a posteriori tree diagram and probability calculations for this example can be seen in Figs. 8-14 and 8-15. Again, ignore or adjust rounding errors at the second and third decimals.

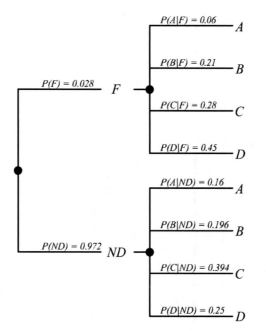

Fig. 8-14. A posteriori tree diagram for Example 2

8.7 Exercises

This section includes three simple sample exercises that should help the reader better understand Bayes' theorem through application and practice. The exercises provide steps for finding the solution to the problem.

Problem 1: Height versus Longevity

According to recent reports, shorter, smaller people tend to have longer lifespans than taller, larger people. In a study involving Japanese-American men, the shorter men carried a "longevity gene" that increased their lifespan. In addition, taller, larger people's bodies demand a higher caloric intake that may predispose them to an increased risk of cancer and possible premature death.

The study classified as short those men whose height was less than 5.18 ft (1.58 m), whereas the tall men were those with a height of greater than 5.41 ft (1.65 m). Of all the men who took part in the study ($n = 8,003$), 4,756 men lived to more than 80 years of age; 1,171 were considered to be short, and 1,244 were considered to be tall, and the rest (2,341 of the men) were in between. Furthermore, the study showed that the probability of surviving past age 90 was 29% for the short men and 16% for the tall men (He et al. 2014).

Considering only the short and tall men (n=1,171+1,244 = 2,415), what is the probability that one of the short men will not live past 90 years of age?

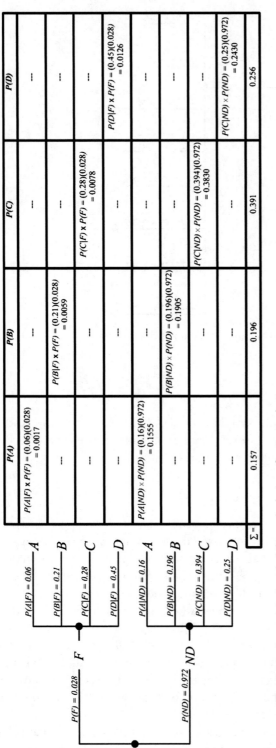

	P(A)	P(B)	P(C)	P(D)
A	$P(A\|F) \times P(F) = (0.06)(0.028)$ $= 0.0017$			
B	--	$P(B\|F) \times P(F) = (0.21)(0.028)$ $= 0.0059$		
C	--	--	$P(C\|F) \times P(F) = (0.28)(0.028)$ $= 0.0078$	
D	--	--		$P(D\|F) \times P(F) = (0.45)(0.028)$ $= 0.0126$
A	$P(A\|ND) \times P(ND) = (0.16)(0.972)$ $= 0.1555$			
B	--	$P(B\|ND) \times P(ND) = (0.196)(0.972)$ $= 0.1905$		
C	--	--	$P(C\|ND) \times P(ND) = (0.394)(0.972)$ $= 0.3830$	
D	--	--	--	$P(C\|ND) \times P(ND) = (0.25)(0.972)$ $= 0.2430$
$\Sigma =$	0.157	0.196	0.391	0.256

$P(A\|F) = 0.06$ A
$P(B\|F) = 0.21$ B
$P(C\|F) = 0.28$ C
$P(D\|F) = 0.45$ D
$P(A\|ND) = 0.16$ A
$P(B\|ND) = 0.196$ B
$P(C\|ND) = 0.394$ C
$P(D\|ND) = 0.25$ D

$P(F) = 0.028$ F

$P(ND) = 0.972$ ND

Fig. 8-15. A posteriori probability calculations for Example 2

Solution:

1) Assign a letter to each event:
   - S = Short
   - T = Tall
   - D = Do not live past 90
   - L = Live past 90
2) Next, determine the a priori and conditional probabilities:

Total number of men considered (tall and short only, older than 80 years) = 2,415

$$P(S) = \frac{1,171}{2,415} = 0.485$$

$$P(T) = \frac{1,244}{2,415} = 0.515$$

$$P(L|S) = 0.29$$

$$P(L|T) = 0.16$$

3) The tree diagram can be constructed as in Fig. 8-16:
4) The joint probabilities can be tabulated. The a priori is calculated by taking the sum of all the joint probabilities in its respective column. This can be seen in the last row of the table in Fig. 8-17.
5) Apply Bayes' theorem to determine the final answer:

$$P(S|D) = \frac{P(D|S) \cdot P(S)}{P(D)} = \frac{0.344}{0.777} = 0.443 = \mathbf{44.3\%}$$

The probability that a short man will not live past 90 years of age is 44.3%

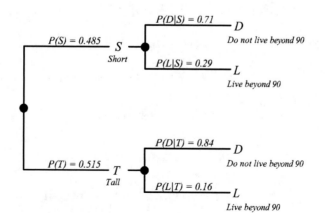

Fig. 8-16. A priori tree diagram for Problem 1

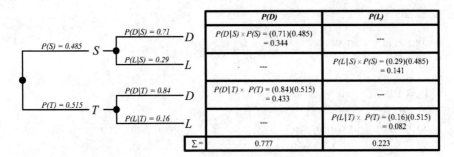

		P(D)	P(L)
$P(D|S) = 0.71$ — D		$P(D|S) \times P(S) = (0.71)(0.485)$ $= 0.344$	---
$P(L|S) = 0.29$ — L		---	$P(L|S) \times P(S) = (0.29)(0.485)$ $= 0.141$
$P(D|T) = 0.84$ — D		$P(D|T) \times P(T) = (0.84)(0.515)$ $= 0.433$	---
$P(L|T) = 0.16$ — L		---	$P(L|T) \times P(T) = (0.16)(0.515)$ $= 0.082$
$\Sigma =$		0.777	0.223

Fig. 8-17. A priori probability calculations for Problem 1

Problem 2: Hurricane Season

Hurricane season lasts for just under four months of the year in the Hawaiian Islands. This season, there is a 15% probability that a hurricane will make landfall. If there is a hurricane, the probabilities of rainstorms or strong winds are 49% each. However, if no hurricane comes ashore this season, the probabilities of rainstorms and strong winds are 25% and 15%, respectively.

What is the probability that there will be a rainstorm during these six months? What is the probability of no hurricane occurring, given that there will be rain storms?

Solution:

1) Assign a letter to each event:
   - H = Hurricane hits
   - N = No hurricane
   - R = Rainstorms
   - W = Strong winds
   - N' = No severe weather
2) Next, determine the a priori conditional probabilities:

$$P(H) = 0.15 \, (\text{Given})$$

$$P(N) = 1 - 0.15 = 0.85 \, (\text{Calculated})$$

$$P(R|H) = 0.49 \, (\text{Given})$$

$$P(W|H) = 0.49 \, (\text{Given})$$

$$P(N'|H) = 1 - 0.98 = 0.02 \, (\text{Calculated})$$

$$P(R|N) = 0.25 \, (\text{Given})$$

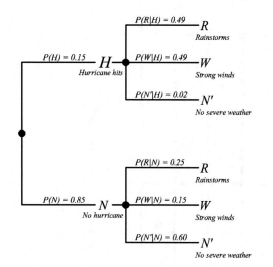

Fig. 8-18. A priori tree diagram for Problem 2

$$P(W|N) = 0.15 \text{ (Given)}$$

$$P(N'|N) = 1 - 0.40 = 0.60 \text{ (Calculated)}$$

3) The tree diagram can now be constructed as shown in Fig. 8-18.
4) Tabulate the joint probabilities. The a priori probabilities of rain, strong winds, and no severe weather are calculated by taking the sum of all the joint probabilities in their columns, as can be seen in the last row of the table in Fig. 8-19.
5) Thus, the probability of a rainstorm this hurricane season is

$$P(R) = 0.286 = \textbf{28.6\%}$$

6) Finally, Apply Bayes' theorem to determine the probability of no hurricane given there will be rain:

$$P(N|R) = \frac{P(R|N) \cdot P(N)}{P(R)} = \frac{0.213}{0.286} = 0.443 = \textbf{0.745}$$

The chance that there will be a rainstorm during the six months of the hurricane season is 28.6%, and the chance that there will be no hurricane, given that it will rain, is 74.5%.

Problem 3: Performance-Enhancing Drugs in Mixed Martial Arts

Performance-enhancing drugs (PEDs) have been used in sports for quite some time. Controversy about their use in mixed martial arts has been brought up on many occasions owing to tests that give inaccurate readings or false positives.

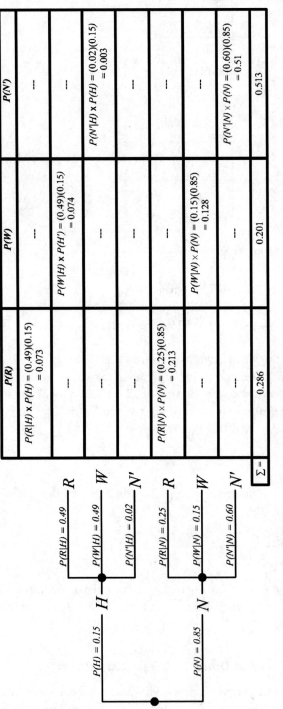

Fig. 8-19. A priori probability calculations for Problem 2

Anabolic steroids have been the standard drug of choice for enhancing an athlete's performance. Because steroids have been in use for quite some time, the tests for them have been quite accurate in detecting whether an athlete has been using this performance enhancer. However, newer technologies have been implemented, and newer, more complicated "drugs" are emerging and being used.

These new performance enhancers are not always drugs per se; instead, athletes are abusing hormone therapies that can be legitimately used to treat hormone deficiencies. Human growth hormone (HGH) and testosterone replacement therapy (TRT) are the most popular at the moment. The problem with testing for abuse of these therapies is that there are instances of false positives owing to spikes in these natural occurring hormones in the test subjects.

Say that there are about 1,000 mixed martial artists in a sample, both professional and amateur, and that 20% of them are actual users of PEDs, both established and new. The tests are 99% accurate in detecting the abuse of hormone therapy if the user is indeed using a PED. Conversely, the tests have a 5% false-positive rate, meaning that tests can return positive results even for users who are not on any hormone therapy or PED.

What is the probability that an athlete not using HGH or TRT will receive a false positive?

Solution:

1) Assign a letter to each event:
   - U = Actual user of PEDs
   - N = Non-user
   - P = Positive result
   - P' = Negative result
2) Next, determine the a priori and conditional probabilities:

$$P(U) = 0.20 \, (\text{Given})$$

$$P(N) = 0.80 \, (\text{Calculated})$$

$$P(P|U) = 0.99 \, (\text{Given})$$

$$P(P'|U) = 1 - 0.99 = 0.01 \, (\text{Calculated})$$

$$P(P|N) = 0.05 \, (\text{Given})$$

$$P(P'|N) = 1 - 0.05 = 0.95 \, (\text{Calculated})$$

3) Now construct the tree diagram (Fig. 8-20).

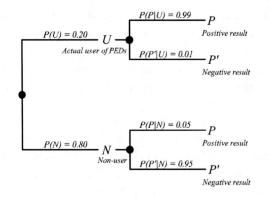

Fig. 8-20. A priori tree diagram for Problem 3

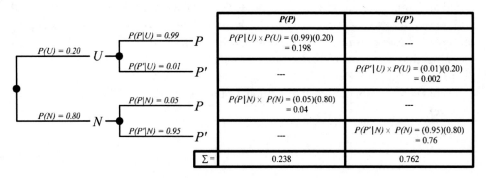

Fig. 8-21. A priori probability calculations for Problem 3

4) Tabulate the joint probabilities. The a priori is calculated by taking the sum of all the joint probabilities in its column. This can be seen in the last row of the table in Fig. 8-21, where the probability of martial artists being tested positive, whether rightly or falsely, is 23.8%.

5) Finally, apply Bayes' theorem to determine the final answer:

$$P(N|P) = \frac{P(P|N) \cdot P(N)}{P(P)} = \frac{0.04}{0.238} = \mathbf{0.168}$$

The probability that an athlete who is not on PEDs and receives a false positive result is 16.8%.

8.8 Conclusion

We derived Bayes' theorem from conditional probability through the understanding of events and sample space. We worked through sample applications and exercises to further demonstrate applications of Bayes' theorem which should help readers better understand the theorem through practice.

This chapter on Bayes' theorem and its applications through the examples demonstrates that the theorem is an excellent tool to determine, or quantify, the probability that a posterior event will occur. Furthermore, we have seen how useful the theorem is for risk analysis enabling data for decision making.

References

Ahmed, A., Kusumo, R., Savci, S., Kayis, B., Zhou, M., and Khoo, Y. B. (2005). "Application of analytical hierarchy process and Bayesian belief networks for risk analysis." *Complexity Int.*, **12**(12), 1–3.

Belhouse, D. R. (2001). "The reverend Thomas Bayes FRS: A biography to celebrate the tercentenary of his birth." <http://www2.isye.gatech.edu/~brani/isyebayes/bank/bayes biog.pdf> (Dec. 20, 2016).

Das, B. (1999). "Representing uncertainty using Bayesian networks." *Publication DSTO-TR-0918*, Dept. of Defence, Defence Science and Technology Organisation, Salisbury, Australia.

Freund, J. E., and Williams, F. J. (1977). *Elementary business statistics: The modern approach*, 3rd Ed., Prentice-Hall, Englewood Cliffs, NJ.

He, Q., et al. (2014). "Shorter men live longer: Association of height with longevity and FOXO3 genotype in american men of Japanese ancestry." *PLoS ONE*, **9**(5), e94385.

SAS Institute (2010). "Bayesian analysis: Advantages and disadvantages." <http://support.sas.com/documentation/cdl/en/statug/68162/HTML/default/viewer.htm#statug_introbayes_sect015.htm> (May 21, 2014).

Vanem, E. (2013). *Bayesian hierarchical space-time models with application to significant wave height: Ocean engineering and oceanography*, Vol. **2**, Springer, Berlin.

CHAPTER 9

Matrix Analysis

9.1 Introduction

An engineer's job goes far beyond calculations and formulas; it usually involves making important decisions. Matrix analysis is one tool that engineers use to structure important decisions and problems. This tool allows decision makers to prioritize their needs and then evaluate and rate the possible solutions. It is used under many different names, but the procedure remains the same and can be applied to any number of circumstances.

Matrix analysis is basically a list of alternatives to be evaluated on one axis, subject to a list of criteria on the other axis. Each of these criteria is weighted based on its importance to the solution.

This decision-making model can be applied to almost any problem, whether or not it is related to engineering. The construction of a decision matrix is easily done by following the COWS (criteria, options, weights, and scores) method, which is outlined in Table 9-1.

The COWS method, which will be further explained throughout this chapter, demonstrates the ease of using matrix analysis as a decision-making model. Table 9-2 shows an example of a weighted evaluation matrix with the different segments labeled. This particular matrix is being used to evaluate the different alternatives for constructing a concrete foundation.

Using the format shown in Table 9-2, matrix analysis can be performed for any type of decision process. Its simple set-up and ease of use make this decision-making model ideal for engineering-related decisions.

9.2 Criteria

The first step in the matrix analysis process is to generate a list of criteria that will be used to judge the alternatives. For example, when purchasing a car, the criteria for choosing a model might include safety, aesthetics, and performance. When selecting criteria, it is important to include only those that differentiate among the

Table 9-1. COWS Method

Symbol	Item	Explanation
C	Criteria	Develop a hierarchy of decision criteria
O	Options	Identify options (also called solutions or alternatives)
W	Weights	Assign a weight to each criterion based on its overall importance
S	Scores	Rate each option on a ratio scale by assigning it a score

Table 9-2. Satisfaction Factor and Rank Evaluation Matrix (Concrete Foundation)

	Criteria					
	Environment	Quality	Reliability	Location		
	Assigned weight (Φ)					
	1	6	3	2		
Alternatives/activities	Satisfaction factors (SF)				$\Sigma(\Phi \times SF)$	Ranking
Fabricate panels	6	1	7	4	41	3
Erect panels	1	3	6	0	37	4
Construction forms	4	5	8	0	54	2
Cast concrete	8	2	9	7	61	1

alternatives. A criterion that assigns the same score to each alternative is really a requirement, and it should not be included in the analysis because it does not contribute anything to the decision-making process. One should also avoid using synonyms as criteria, because they unnecessarily duplicate criteria already accounted for. For example, considering "maneuverability" and "turning radius" as separate criterion in the car example is likely to include tremendous overlap. Further, it is important to consider only criteria that lead to subjective decisions.

9.3 Alternatives/Activities

The next step is to identify the options or alternatives/activities—in other words, the items being evaluated and selected. Because a large number of alternatives/activities (henceforth just *alternatives*) are generally available, it is important to develop strategies for narrowing them down. One method is to screen out alternatives that fail to satisfy mandatory requirements, so that the only alternatives evaluated are those with subjective and/or optional criteria. For example, when deciding among cars, a requirement could be that the car must have four doors, thereby eliminating two-door cars from the evaluation.

Many information sources, such as personal knowledge and technical literature, are available for researching and selecting alternatives. Previously completed engineering projects serve as a common source of input for alternatives. Looking at completed projects makes it easier to see what points of emphasis the project focused on; it can also help a decision maker avoid the failed aspects of a previous project.

Generally speaking, it is best to evaluate three or more alternatives; two alternatives do not provide enough options. When analyzing three or more alternatives, it is easier to see the flaws in some alternatives and the benefits of others. However, the number of alternatives depends on the number of good choices available. Sometimes that number may be two, whereas at other times it may be a few dozen.

9.4 Weights

Next, we assign a weight to each criterion based upon its overall importance in the decision-making process. One can arbitrarily assign weights to the criteria based upon feelings or take a more methodical approach. Such an approach evaluates each criterion independently, allowing for a ranking of importance for criteria. Using these scores, one can easily establish values for the weights in the evaluation matrix.

The analytic hierarchy process (AHP) provides one way to assign weights systematically, removing subjectivity from that process. This method will be covered in depth in its own section in Chapter 10.

In matrix analysis, we must create a matrix that compares the criteria independently (Kaufman 1998). Table 9-3 demonstrates a preference matrix, using the example of the construction of a concrete column. We will be analyzing the criteria of initial cost, safety, reliability, and maintainability.

Each box in the preference matrix shows a comparison between two criteria. Using the importance scale on the right, we can rank our preference for each individual criterion against other criteria. For example, the top left box in Table 9-3 shows a comparison between criterion A (initial cost) and criterion B (safety). The user in this case greatly prioritizes initial cost (A) over safety (B). Hence, the user assigns a score of A3, signaling that A has a major preference (a score of 3) compared

Table 9-3. Preference Matrix

	B	C	D	Importance scale	Alternatives
A	A3	C1	A3	3 – Major preference	A – Initial cost
	B	C3	B2	2 – Medium preference	B – Safety
		C	C2	1 – Minor preference	C – Reliability
			D		D – Maintainability

to B. This process is then repeated to complete the preference matrix, giving us an idea of how each criterion stacks up when compared to the others. This process leads to a less biased, less random, weight assessment and, consequently, a better decision.

Now all the preferences on the importance scale for each alternative are added to find its total raw score. Because A twice received a major preference, the raw score of A is 6. Because B was preferred only once, with a medium preference, its raw score is 2. Similarly, the raw score of C is 6, and that of D is 0. Once all preferences have been obtained, the user tallies the scores into a weight evaluation matrix, as illustrated in Table 9-4.

Using the raw scores, a user can create a weight system using a scale of his or her choice. For example, this user decided on a standard 10-point scale. This is usually a good step for those who like percentages associated with their weights, but it is not necessary; one can simply use the raw scores obtained from the preference matrix. In this case, maintainability can be excluded from the list because it received no preferred score. However, we will leave it in this analysis for illustration. The assigned weights can now be imported into the evaluation matrix, as outlined in the next section.

9.5 Scores

Each alternative can now be evaluated and scored for satisfaction based upon each criterion. These scores, called satisfaction factors, measure how important the user thinks it is that the alternative satisfy the different criteria. To keep things simple, it is generally best to use a scale of 1 to 10 for satisfaction factors, but any numerical scale that the user prefers can be used. Table 9-5 presents an example of an evaluation matrix, continuing the example of the construction of a concrete column. We have now imported the assigned weights into this evaluation matrix and can assign satisfaction factors to the alternatives.

To complete the evaluation and make a decision, the user must fill in the satisfaction factors. We will assign a score to each alternative and criterion based on our chosen scale—in this case, from 1 to 10. The top left cell, for example, asks the

Table 9-4. Weight Evaluation Matrix

Construction of a concrete column		Determining weights	
		Raw score	Assigned weight (Φ)
A	Initial cost	6	4.3
B	Safety	2	1.4
C	Reliability	6	4.3
D	Maintainability	0	0
		14	10	Total

Table 9-5. Satisfaction Factor and Rank Evaluation Matrix for the Construction of a Concrete Column

	Initial cost	Safety	Reliability	Maintainability		
	Assigned weight (Φ)					
	4.3	1.4	4.3	0		
Alternatives/activities	Satisfaction factors (SF)				Σ(Φ × SF)	Ranking
Fabricate panels	7	1	4	3	48.7	3
Erect panels	1	2	7	0	37.2	4
Construction forms	3	4	8	0	52.9	2
Cast concrete	9	2	8	9	75.9	1

user to rank the fabrication of panels with regard to initial cost. The user decided that it deserves a satisfaction score of 7 out of 10, which indicates that it is relatively important.

The satisfaction factors section helps the user decide which criteria apply most to each alternative. Weights are also a large factor because they are multiplied by the satisfaction factors to obtain the final score. Thus, even though the alternative of cast concrete received a high satisfaction factor for maintainability, that does not affect its score because maintainability has been assigned a weight of 0. Once all the scores are summed, we can rank the alternatives from highest to lowest score. For this particular project, the alternative of cast concrete has the highest score (75.9) and will receive the most attention from the construction managers (Miles 1989; Borysowich 2006).

9.6 Case Study 1: Construction Site Preparation

In this case study, we will apply matrix analysis to site preparation for construction. This process usually involves preliminary surveying of the site, trench excavation, earth backfill, and running overhead lines. The use of matrix analysis can allow a construction management company to focus on particular steps—such as the site survey—that may be more important to the completion of the project than the other steps. We will begin by determining the criteria to be used in the analysis. For this example, we will examine the site's environment, quality, reliability, and location.

When analyzing the environment (A) as it pertains to site preparation, we do not consider it extremely significant when compared to site quality (B) and reliability (C). Although the environment is important, a high-quality site will do more to allow for easier construction. Hence, in Table 9-6 we see that when environment (A) is compared to quality (B), reliability (C), and location (D), it gets scores of B3, C3, and A1, respectively.

Table 9-6. Preference Matrix for Site Preparation: Case Study 1

	B	C	D		Importance scale	Alternatives
A	B3	C3	A1		3 – Major preference	A – Environment
	B	B1	B2		2 – Medium preference	B – Quality
		C	D2		1 – Minor preference	C – Reliability
			D			D – Location

Table 9-7. Weight Evaluation Matrix for Site Preparation: Case Study 1

Site preparation		Determining weights		
		Raw score	Assigned weight (Φ)	
A	Environment	1	0.83	
B	Quality	6	5.0	
C	Reliability	3	2.5	
D	Location	2	1.67	
		12	10	Total

We have decided in this case to assign the weights on a scale of 1 to 10 for ease of use. As can be seen in Table 9-7, normalizing the scores to that scale results in weights of 0.83 for environment, 5 for quality, 2.5 for reliability, and 1.67 for location. Emphasis is placed on the quality and reliability of the site, rather than the environment, because of the need for good customer service. Nothing can delay a project and increase the price like a bad site with irregular soil, so we choose to emphasize those criteria, as shown in Table 9-7.

As before, we assign satisfaction scores (see Table 9-8) (Miles 1989; Borysowich 2006). The evaluation matrix (Table 9-8) reveals that site survey received a rank of 1 and therefore requires the greatest focus from the construction manager. It is followed in importance by the overhead line, which earned a rank of 2.

The construction management firm or manager will choose the site survey and overhead line as the focal points of site preparation, with the site survey taking top priority. This information can help a manager make better decisions regarding personnel deployment, materials procurement, and construction scheduling.

Most construction managers would choose to perform this simple evaluation in their heads or rely on previous work. However, outlining the alternatives and weights using the matrix analysis model can enable more informed decisions, which will lead to a more well-founded and successful construction project.

Table 9-8. Satisfaction Factor and Rank Evaluation Matrix for Site Preparation: Case Study 1

Alternatives/activities	Environment	Quality	Reliability	Location	$\Sigma(\Phi \times SF)$	Ranking
		Assigned weight (Φ)				
	0.83	5.0	2.5	1.67		
		Satisfaction factors (SF)				
Site survey	1	9	9	9	83.4	1
Excavate trench	8	2	3	4	30.8	4
Backfill earth	7	4	2	4	37.5	3
Overhead line	1	8	9	8	76.7	2

9.7 Case Study 2: Selection of an Engineering College

A very important step in an engineer's career is selecting a college, which can involve very confusing decisions because of the variety of schools to choose from throughout the nation. Everyone has different criteria for selecting a school. One person may consider only schools located in warm climates, whereas another might heavily emphasize education quality. Some choose schools with strong athletic programs, and others prefer schools that are close to home. Ultimately, selecting a college is a decision involving a lot of alternatives and criteria. Using matrix analysis can help a student make a more informed decision about his or her future education.

As a second case study, we will use a student who is selecting an undergraduate school. During the selection process, the student chooses to emphasize proximity to home (A), academic quality (B), school size (C), and college life (D). These may differ from other students' criteria, but this student will base the decision upon them, as outlined in Table 9-9. Cost may be another important criterion but not so in this case.

As further seen in Table 9-10, the student heavily emphasizes colleges with high academic standards (B) and exciting student life (D). However, having the school close to home (A) and incurring a lower cost (C) are less important. Despite A and C having low weights, they still have an effect in determining where the student ends

Table 9-9. Preference Matrix for College Selection: Case Study 2

	B	C	D	Importance scale	Alternatives
A	B3	A1	D3	3 –Major preference	A – Proximity to home
	B	B2	D3	2 – Medium preference	B – Academic quality
		C	C1	1– Minor preference	C – Cost
			D		D – College life

Table 9-10. Weight Evaluation Matrix for College Selection: Case Study 2

Site preparation		Determining weights		
		Raw score	Assigned weight (Φ)	
A	Proximity to home	1	0.77	
B	Academic quality	5	3.85	
C	Cost	1	0.77	
D	Student life	6	4.61	
		13	10	Total

up. Even one extra point can tilt a decision one way or another. Now, using these weights, the student can evaluate the college choices using the desired criteria.

As in the previous example, the raw score represents the weight of each criterion on a scale of 10. The raw scores for all criteria total 13 points, and then the assigned weight normalizes each raw score on the scale of 10. Thus, the assigned weight for academic quality is $5/13 \times 10 = 3.846$, rounded to 3.85.

It needs to be noted that Table 9-10 only evaluates the relative importance of the criteria, establishing a weight for each when compared to the other criteria. In Table 9-11, we multiply the weights by the satisfaction factors that represent how well each alternative being evaluated fulfills each criterion. Satisfaction factors and weights are different things and must not be confused for one another.

According to the satisfaction factor and rank evaluation matrix in Table 9-11, University A satisfies the most criteria, and University C is not far behind in the ranking. The choice between the two schools is very close, and it only takes a minuscule item to sway the decision. This situation is a perfect demonstration that every satisfaction factor counts, even if its assigned weight is relatively low.

Prior to this, we had applied matrix analysis only to engineering. Using this tool for selecting a school shows that matrix analysis can be used for any decision, in any

Table 9-11. Satisfaction Factor and Rank Evaluation Matrix for College Selection: Case Study 2

Alternatives/activities	Proximity to home	Academic quality	Cost	Student life	$\Sigma(\Phi \times SF)$	Ranking
	Assigned weight (Φ)					
	0.77	3.85	0.77	4.61		
	Satisfaction factors (SF)					
University A	7	9	5	10	90	1
University B	10	7	3	8	73.8	2
University C	5	9	10	2	55.4	3
University D	1	9	1	4	54.6	4

discipline or genre. The ranking obtained from matrix analysis does not have to be the final answer, but can help the user make a more informed decision. Most significantly, we do not simply assign weights to the criteria; we then combine the weights with the satisfaction factors for a complete analysis.

9.8 Case Study 3: Selecting Solar Panels

For the third case study, we will return to an engineering problem in the construction industry. One of the very important decisions to be made during construction is materials procurement. It is sometimes necessary to make sure that elements are of high quality, whereas in other cases this aspect can be overlooked. For example, it would be wise to focus spending on important and expensive items such as the structure of a house, while skimping on light fixtures. Matrix analysis is perfect for deciding among different brands of the same material. For instance, to choose among four different front doors, we can outline the different brands and subject them to the criteria that we have determined are important for the project.

This case study focuses on solar panel procurement for a housing project. When procuring materials, it is generally important to focus on cost, warranty, reliability, and product efficiency. These criteria, however, are all arbitrary, and can depend on the construction manager and the architect's vision. The matrix can be altered to fit each managing party's creative wishes.

Let's now look at the preferences among the different criteria for procuring solar panels. The preferences are established in Table 9-12; that information is then taken and presented in Table 9-13, where we see that there is a more even spread of weights among the four criteria than we saw in the previous case studies. Based on the raw scores, we see that the efficiency of the solar panels is the most important criterion to be considered, because it scores a 4 out of 10 whereas the other criteria score lower. Thus, the warranty for the solar panel is not nearly as important as its efficiency. Just as in the previous case, every point counts and could be the difference between one ranked order and another.

Now let's look at the satisfaction factors for the four different solar panel alternatives in Table 9-14. Brand B satisfies cost perfectly, whereas Brand C satisfies warranty perfectly. But as can be expected from the high warranty score, Brand C can barely satisfy the cost criterion. Although Brand D does not have a high warranty

Table 9-12. Preference Matrix for Solar Panels Procurement: Case Study 3

	B	C	D		Importance scale	Alternatives
A	A1	C1	D1		3 – Major preference	A – Cost
	B	B1	D2		2 – Medium preference	B – Warranty
		C	C2		1 – Minor preference	C – Reliability
			D			D – Efficiency

Table 9-13. Weight Evaluation Matrix for Solar Panels Procurement: Case Study 3

Solar panel procurement		Determining weights		
		Raw score	Assigned weight (Φ)	
Brand A	Cost	2	2	
Brand B	Warranty	1	1	
Brand C	Reliability	3	3	
Brand D	Efficiency	4	4	
		10	10	Total

Table 9-14. Satisfaction Factor and Rank Evaluation Matrix for Solar Panels Procurement: Case Study 3

	Cost	Warranty	Reliability	Efficiency		
	Assigned weight (Φ)					
	2	1	3	4		
Alternatives/activities	Satisfaction factors (SF)				$\Sigma(\Phi \times SF)$	Ranking
Brand A	7	3	6	7	63	2
Brand B	10	4	8	3	60	3
Brand C	1	10	7	9	69	1
Brand D	2	5	4	7	49	4

score, its cost satisfaction is also not high, indicating that it is quite expensive; therefore, in Brand D we have a somewhat unreliable product that is also high priced. According to the satisfaction factors laid out by the construction management firm and reflected in the rankings of Table 9-14, it turns out that the Brand C panels have the highest rank, mostly because of their high score in efficiency. Brand C received the most points despite being the most expensive solar panel considered. However, the Brand A solar panel was a close second and could be considered should any circumstances change for the project. For example, if the project faces a financial shortfall, we could end up using the Brand A panels as a fallback plan, because of their relatively low cost and overall high performance in the procurement matrix analysis.

9.9 Case Study 4: Bid Analysis

Construction management firms always face decisions about which projects to submit bids for. Sometimes a bid, despite its high chance for profitability, might require too much work to place relative to its profit probability. This is where matrix analysis can help a construction management firm outline its alternatives and determine where it should focus its efforts. Time is money, and any time that is not spent on an unnecessary bid is time that can be used for a more worthwhile project.

This case study focuses on four different projects on which a contractor is considering bidding. The contractor may not be able to bid on all projects for want of time and resources. At this time, the evaluation criteria for this particular construction management firm are profitability, time to bid, distance from its office, and workforce required. This is a relatively simplified simulation, because construction cost estimation usually must account for many more criteria. Tables 9-15 and 9-16 show the preferences and weights. By now, students should be well versed in what the preferences mean and how the weights (raw scores and normalized scores) are calculated.

For most construction management firms, the potential profit from a project often takes preference when compared with other factors. The purpose of a construction management firm is to make a profit, and more profits mean potential future jobs. The time to bid is also an important factor because no firm wants to waste time bidding on a job with too low a chance of winning it.

Now we can import the assigned weights into the evaluation matrix, shown in Table 9-17. The highway construction project turns out to be the highest-ranked alternative because of the high weight assigned to profitability. However, the bridge foundation project also has a high potential profit and a relatively high score in the

Table 9-15. Preference Matrix for Bid Analysis: Case Study 4

	B	C	D	Importance scale	Alternatives
A	A2	A2	A3	3 – Major preference	A – Profitability
	B	C1	B2	2 – Medium preference	B – Time to bid
		C	D1	1 – Minor preference	C – Distance from office
			D		D – Workforce required

Table 9-16. Weight Evaluation Matrix for Bid Analysis: Case Study 4

Bid analysis		Determining weights		
		Raw score	Assigned weight (Φ)	
A	Profitability	6	6	
B	Time to bid	2	2	
C	Distance from office	1	1	
D	Workforce required	1	1	
		10	10	Total

Table 9-17. Satisfaction Factor and Rank Evaluation Matrix for Bid Analysis: Case Study 4

Alternatives/activities	Profitability	Time to bid	Distance from office	Workforce required	$\Sigma(\Phi \times SF)$	Ranking
		Assigned weight (Φ)				
	6	2	1	1		
	Satisfaction factors (SF)					
Bridge foundation	8	5	3	3	64	2
Highway construction	10	2	5	1	70	1
Underground trench	3	9	8	5	49	3
Pipe laying	2	10	2	8	42	4

matrix analysis. If the firm cannot assemble a workforce large enough for the highway construction project, it might decide to bid on the bridge foundation project instead. There are always other factors that can be considered, but matrix analysis provides a basis for making an informed decision.

9.10 Discussion

In all facets of life, people have to deal with many alternative decisions and associated criteria. Usually, people decide using a minimalist approach. However, using matrix analysis is an easy way to formulate one's decision-making process while considering more criteria. First, we determine the criteria applicable to the situation, and then we evaluate them to establish a precise weighting system. Next, we apply these criteria and their assigned weights to the alternatives themselves, allowing us to evaluate our alternatives based on how well they satisfy our criteria. Thus, matrix analysis is another way to gain additional information when making a decision. It does not have to directly determine the solution to the problem, but it can help with formulating ideas and turning them into concrete decisions.

9.11 Exercises

Problem 1

You have to choose among four fine-dining restaurants for dinner on your wedding anniversary. This had better be a good selection! Identify four to five criteria for selection and do a pairwise comparison by making a preference matrix to determine normalized weight preferences among those criteria. Undertake satisfaction analysis, and find the rank order of the preferred restaurants.

Problem 2

Your construction company needs to buy a new medium-sized bulldozer for its emerging earthwork projects. Among the criteria the company is interested in are cost, reliability, warranty, maintenance expenses, gasoline consumption, horsepower, and capability of performing automated operations. Prepare a preference matrix for these criteria, find the raw weights and normalized weights, undertake a satisfaction factor analysis, and determine the priority ranking among the three companies that offer bulldozers in the size you need.

References

Borysowich, C. (2006). "Alternatives evaluation matrix." <http://it.toolbox.com/blogs/enterprise-solutions/alternatives-evaluation-matrix-13138> (Apr. 3, 2010).

Kaufman, J. J. (1998). *Value management: Creating competitive advantage*, Crisp Learning, Menlo Park, CA.

Miles, L. D. (1989). *Techniques of value analysis and engineering*, 3rd Ed., Lawrence D. Miles Value Foundation, Washington DC.

CHAPTER 10

MCDM and the Analytic Hierarchy Process

10.1 Introduction

The *analytic hierarchy process* (AHP) was first introduced by Saaty (1977) and is now a widely accepted method in *multicriteria decision making* (MCDM). AHP uses an approach that is subjective but also mathematical and analytic, assigning weights to preferences and consequently arriving at a final decision on the viability of those weights. Once a problem is selected and performance factors (or decision criteria, design factors, etc.) are identified, one can begin comparing those factors and criteria with one another according to their importance.

What was once an art for evaluating weights has been transformed by MCDM and AHP into a science in which rules, techniques, and guidelines are formulated to arrive at a decision. The weighting process can be treated as a scientific way to meet people's requirements and preferences. Consequently, AHP facilitates potential decisions or factors to be analyzed so that decision makers can reach the most desirable and sustainable choices. There are a variety of different tools one can use in decision sciences, but this chapter focuses on the analytic hierarchy process for analyzing the weights assigned to criteria. The AHP method uses a very mechanical approach, which is one of the reasons that it is a method of choice in engineering.

10.2 The Analytic Hierarchy Process (AHP)

AHP uses paired comparisons, mathematical procedures, and the subjective judgments of professionals to make a decision. Multicriteria decision making is relevant to many complex, multivariate real-life issues, which is why learning this process is so helpful. The overall method is called *multicriteria decision analysis* (MCDA), which encompasses "mathematics, management, informatics, psychology, social science, and economics" (Ishizaka and Nemery 2013), creating a stepping-stone toward arriving at a compromise solution. In essence, and traditionally, MCDA helps to optimize multiple criteria that are in play, in order to optimize the net output of the system. Saaty (1982) incorporated the previously used and accepted methods of hierarchical structuring, pairwise comparisons, judgments, eigenvectors, and consistency checks, which proved to be quite useful. Many researchers take interest in AHP because of the objective, mathematical approach it provides compared to traditional techniques, as well as for the relatively easily obtained data it requires for input.

AHP modeling comes in quite handy in group decision-making processes. The numerical value given to each alternative can be derived from the input of many group members. It offers a chance for all group members to express their thoughts on the problem and to reach a conclusion effectively. Group decision making helps to keep judgments about an issue normalized and consistent. Hence, it helps to increase the reliability of AHP because each member has the chance to authenticate or verify what the other members are thinking, and outlandish or aberrational weight assignments can be intercepted.

Individuals or groups will brainstorm a variety of alternatives and corresponding desired criteria, both of which are subjective. The brainstormed list may not cover all possible criteria and alternatives, but the set of alternatives and criteria it includes will be suitable for the decision maker. By the rules of brainstorming, there are no right or wrong answers, and different individuals will create different lists.

The goal of AHP is to rationalize individual judgments by assigning values to them. The values are numbers within certain ranges and can differ from criterion to criterion; later they will be normalized in order to properly compare all the alternatives and criteria. Factors that are similar in type may be combined, and other factors may be broken down in more detail.

Let us examine the name of this method. The word *analytic* is, of course, a form of the word *analysis*, which means a breaking down into parts; a *hierarchy* is a set-up or list that is prioritized by importance; and a *process* is a series of actions, changes, or functions that bring an end result. Together, AHP offers a systematic and structured methodology for analyzing decisions that depend on weight assignments.

Pairwise comparisons are initially used to derive the pertinent data used in the decision-making process. (You may recall that paired analysis was discussed in Chapter 9.) The paired comparisons are then evaluated based on the importance of each criterion. From there, each alternative is analyzed. At the end of the mathematical process, the best alternative is determined. AHP may not necessarily find the *right* answer, but it finds the *best* answer based on the data available (Forman and Selly 2001).

10.3 Users of AHP

According to Saaty (1982), AHP is useful to individuals who must make decisions and set priorities. This describes most people who are involved in everyday decision making at work or home, whether for official or personal reasons. Often, daily and long-term activities do not allow an individual to analyze a situation at such a microscopic level, but each individual is able to take a look at an overall situation and come to a conclusion. AHP uses a simple mathematical approach to analyze both simple and complex decisions.

10.4 AHP Approaches—The Deductive Approach and the Systems Approach

Individuals generally arrive at a decision by using two approaches that represent how they think: the deductive approach and the systems approach.

In the *deductive* approach, systems can be analyzed by representing them as networks of structures and chains. For example, individuals can break down a network, look for explanations of the function of each part, and then determine an explanation of why the network exists. This approach to decision making, though vitally useful, may result in individuals failing to see the relationships among all the parts, as well as items that do not appear in the structure.

The *systems* approach, conversely, examines the entire system from a general perspective in order to make the best decision. Systems theorists believe that examining the system as a whole, rather than the function of each part, can give us a better understanding of the function of the entire matter at hand. An example would be evaluating how a car runs, how it interacts with other cars on the road, how it reacts to external elements, and so forth, ignoring the fact that the car is composed of thousands of smaller parts, which have their own subsytems.

The ideal situation when trying to arrive at a final decision is to combine the two approaches. Setting up the hierarchy, using the deductive approach, and then stepping back and taking a systematic approach will obviously be the best decision route.

10.5 Weighting Criteria

Many potential applications for assigning weights to decision parameters can be found in construction. For instance, behavioral parameters may be evaluated, such as in a partnering example presented in Singh (2005). Other potential applications involve evaluating traditional engineering performance factors such as reliability, safety, comfort, and quality. One such application could be for a general contractor compiling a bid proposal, in which the subcontractor offering the lowest bid may not necessarily be the most qualified. Using AHP to analyze and prequalify subcontractors can consequently help the general contractor put together a qualitative measure of which subcontractor is ideal to use on the project. Ruling out incompetent subcontractors will result in a smoother project with fewer problems.

Table 10-1. Goal and Criteria Requirements

Goal	Criteria
Find competent subcontractors	Work quality
	Communication
	Reliability
	Reputation
	Respectability

Table 10-1 shows the decision maker's brainstorming process for deciding which criteria are required to arrive at the ultimate goal. The goal is to find a competent subcontractor to complete the project, which encompasses a variety of criteria. The ideal subcontractor must have good work quality, communication, reliability, and reputation. However, each individual decision maker weighs these criteria differently and has a different idea about what is important. The subjective reasoning behind each weight will make a difference in determining the ideal alternative.

10.6 Structuring a Problem

As the *H* in the acronym indicates, AHP is structured according to a hierarchy based on judgments; this hierarchy is generally rendered as a pyramid. The top level is the *goal.* In this specific example, the goal is to find the ideal subcontractor. The second level represents the *alternatives.* In this example there are three alternatives: Subcontractor A, Subcontractor B, and Subcontractor C. The lowest level represents the *criteria.* In this example, we will use the criteria of work quality, communication, reliability, reputation, and respectability. It is possible to add more levels and develop more complex hierarchies, but AHP requires a minimum of three levels.

Each level of the hierarchy uses pairwise comparisons to compare the items. This is where using a numerical scale, such as 1–9[1], is useful for comparing the criteria or alternatives. On this scale, 1 represents the least important and 9 the most. Fig. 10-1 shows the hierarchy for this example AHP model. Each subcontractor is rated based on previous history. Work quality, communication, reliability, reputation, and respectability are all weighted and analyzed.

10.7 Cross Tabulation

Cross tabulation involves putting criteria and alternatives in a cross table to solve a multicriteria decision-making process (Teknomo 2006). It is a simple process in

---

[1]Other scales, such as 1–5 or 1–100, are also feasible.

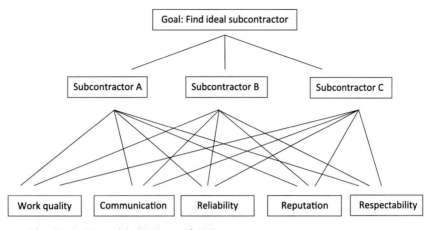

Fig. 10-1. Hierarchy structure of AHP

Table 10-2. General Contractor's Evaluation of Subcontractors Based on Criteria

Criteria | Alternatives	Subcontractor A	Subcontractor B	Subcontractor C	Range
Work quality	8	5	6	1 to 10
Communication	2	5	−1	−5 to +5
Reliability	30	90	40	1 to 100
Reputation	4	3	4	0 to 5
Respectability	7	10	10	0 to 10
Sum	51	113	59	
Normalized score	22.87% (51/223)	50.67% (113/223)	26.46% (59/223)	

Note: Total = 51 + 113 + 59 = 223.

which one lists the criteria in the far left column and the alternatives along the top row. Then the decision maker fills in each cell with subjective values that correspond to the chosen range. The scores are added down each column and then normalized relative to each other. In this case, Table 10-2 shows Subcontractors A, B, and C charted against the five criteria required by the general contractor.[2]

Each criterion has been assigned a different range, for which the general contractor—as decision maker—has to determine the most "accurate" number, which, although it will generally be subjective, must be as representative as possible. Each score should then be normalized so that the comparison is more accurate. In this simple case, the normalized score is simply the ratio of the sum of each subcontractor's scores to the total possible score.

---

[2]Negative scores can also be assigned; note that Subcontractor C has a −1 score for communication.

From this example, it is clear to see that Subcontractor B (50.67%) is a better choice than Subcontractor C (26.46%), which is a better choice than Subcontractor A (22.87%). However, further analysis is required to more certainly find the best alternative.

10.8 Criteria Ranges

10.8.1 Ranges by Rank

Various ranges have been assigned to these criteria because, in real life, there are many dimensions to take into consideration. If the decision maker were simply comparing dollar amounts for similar items, then the range would not show much fluctuation. However, because this is a real-life example involving criteria that are not equivalent, the ranges vary.

The next step is to analyze each subcontractor based on the rank of each factor. The values from the previous table are changed to ranks in Table 10-3, and new choices are entered. The most favorable choice is ranked 1 and the least favorable choice 3; a more favorable choice means a higher score from Table 10-2. Because Subcontractor A ranked highest in work quality, it receives a 1; because Subcontractor B scored the lowest in work quality, it receives a 3.

Because in this case a lower sum represents a better score than a higher sum, the normalized scores in Table 10-3 are calculated by using Eq. (10-1):

$$\text{Normalized Score} = \frac{1}{2}\left(1 - \frac{\text{Sum}}{\text{Total Sum}}\right) \times 100 \qquad (10\text{-}1)$$

Adding up the total scores of 9, 8, and 9 for Subcontractors A, B, and C, respectively, gives us a total of 26. Hence, the normalized score for Subcontractor A is $0.5 \times (1 - (9/26)) = 0.3269$, or 32.69%. The normalized scores for Subcontractors B

Table 10-3. Evaluation of Subcontractors Based on Rank of Each Factor

Criteria │ Alternatives	Subcontractor A	Subcontractor B	Subcontractor C
Work quality	1	3	2
Communication	2	1	3
Reliability	3	1	2
Reputation	1	2	1
Respectability	2	1	1
Sum	9	8	9
Normalized score	32.69%	34.62%	32.69%

Note: Total = 9 + 8 + 9 = 26.

and C are calculated similarly. Subcontractor B again corresponds to the highest preference, with a normalized score of 34.62%. These values are shown in Table 10-3.

10.8.2 Range Between 0 and 1

Next, we will convert the values of Table 10-3 so that each factor has the same range value. This makes for a more realistic and fairer comparison. Remember that Table 10-2 used a variety of different ranges for the criteria, but comparing the criteria with different range values makes for skewed results. To normalize these values, the decision maker can pick any range of values, but this example will use a range of 0 to 1. The formula in Eq. (10-2) is used; it is based on the simple geometry of a line segment.

$$\text{New Score} = \left(\frac{\text{NUB} - \text{NLB}}{\text{OUB} - \text{OLB}}\right) \times (\text{Original Score} - \text{OLB}) + \text{NLB} \qquad (10\text{-}2)$$

where

NUB = new upper boundary,
NLB = new lower boundary,
OUB = old upper boundary, and
OLB = old lower boundary.

For example, the new upper boundary of Subcontractor A's work quality is 1 and the new lower boundary is 0. The old upper and lower boundaries were 10 and 1, with an assigned score of 8. Thus, the converted score, based on the new range, for Subcontractor A's work quality is $[(1 - 0)/(10 - 1)] \times (8 - 1) + 0 = 0.78$. The converted scores are given in Table 10-4.

The normalized scores are calculated in the same way as in Table 10-2. Hence, for Subcontractor C, the normalized score is $3.15/10.36 = 0.3040$, or 30.40%.

Table 10-4. Converted Scores Based on Range

Criteria | Alternatives	Subcontractor A	Subcontractor B	Subcontractor C
Work quality	0.78	0.44	0.56
Communication	0.70	1.00	0.40
Reliability	0.29	0.90	0.39
Reputation	0.80	0.60	0.80
Respectability	0.70	1.00	1.00
Sum	3.27	3.94	3.15
Normalized score	31.56%	38.03%	30.40%

Note: Total = 3.27 + 3.94 + 3.15 = 10.36.

10.8.3 Weighting the Criteria

Now let us assume that one criterion is more important than the others by weighting the importance of each criterion based on the best judgment of the project manager or team. Assigning a relative importance level to each criterion will greatly affect the outcome, because now the criteria are not all equal. This weighting process will help the decision maker in the final step of the decision-making process.

Now, undertake a preference matrix analysis as in Chapter 9. Let's say that reliability and reputation are each two times more important than respectability, communication is two times more important than reliability, and work quality is three times more important than reliability, and so on. This is shown in Table 10-5. Now we calculate each score's importance weight by dividing the criterion's assigned score by the sum of all the scores. For example, work quality's importance weight is calculated as $(6/15) \times 100 = 40\%$.

The next step is to multiply the weight of each criterion by the new, converted scores from Table 10-4. For example, the new work quality value for Subcontractor A is calculated by multiplying 0.78 (the converted score) by 40% (the importance weight). The new values can be found in Table 10-6. Ignore rounding errors.

After factoring in the importance of each criterion and evaluating it against the weighting of subcontractor importance, the normalized scores indicate that Subcontractor B (36.22%) is still the most preferable, showing that the scores are consistent across these methods. It is important to evaluate in depth each criterion

Table 10-5. Weights of Importance

	Work quality	Communication	Reliability	Reputation	Respectability	Sum
Importance level	6	4	2	2	1	15
Importance weight	40.00%	26.67%	13.33%	13.33%	6.67%	

Table 10-6. Weighted Scores

Criteria | Alternatives	Weight	Sub A	Sub B	Sub C	
Work quality	40.00%	0.31 (0.78 × 0.4)	0.18	0.22	
Communication	26.67%	0.19	0.27	0.11	
Reliability	13.33%	0.04	0.12	0.05	
Reputation	13.33%	0.11	0.08	0.11	
Respectability	6.67%	0.05	0.07	0.07	
Sum	100%	0.69	0.71	0.55	1.96
Normalized score		35.20% (0.69/1.96)	36.22%	28.06%	

and alternative in order to ensure that the assigned scores are as accurate as possible. If an inaccurate judgment is made during the first steps, the final result may not be accurate.

Of course, the decision maker should realize that the scores given to each criterion and alternative are partly subjective. Thus, another type of decision-making tool should probably be used as well in order to arrive at the ideal decision.

10.9 Comparison Matrix

Comparison matrices are helpful when faced with multicriteria decision making. To enable this decision-making tool, the decision maker first conducts paired comparisons using best judgments. Each decision maker is able to express his or her individual opinion about a single pairwise comparison at a time. Pairwise comparisons allow for easier expression and evaluation than performing a simultaneous comparison among all alternatives.

10.9.1 Pairwise Comparison

Let us now assume that a project manager is trying to find the most suitable project engineer for a future $30 million project. Based on his best judgment from previous experience, he comes up with a matrix to compare the performance of three different project engineers. The intensity of importance is ranked on a scale of 1–9.

Perhaps the decision maker evaluates the importance of a criterion using a verbal scale. This verbal scale can be converted to a numerical, or Likert, scale by adhering to Saaty's (1980) rules. According to Saaty (1982), a numerical value of 1 in a paired comparison means that both alternatives are equal in importance; both activities contribute equally to the objective at hand. If a value of 3 is chosen, one alternative has a weak importance compared to the other or is slightly favored over the other. If a value of 5 is chosen, one alternative has an essential, or strong importance compared to the other. If a value of 7 is chosen, one alternative has a demonstrated importance compared to the other. If a compromise is needed, then an intermediate value of 2, 4, 6, or 8 may be picked. As may be expected, a smaller scale, such as 1–5, would not offer as much detail as a 1–9 scale, whereas a larger scale, such as 1–100, would make it too hard for the decision maker to properly analyze the situation.

A pairwise comparison, as demonstrated in Fig. 10-2, is used to find the relative importance of each alternative to each required criterion.

The accurate estimation of pertinent data is one of the critical steps in decision making. It is often difficult or impossible to quantify each alternative correctly. The decision maker must use the relative importance or weight of each alternative to aid in decision making. The matrices in Fig. 10-2 show three comparisons among three project engineers. A 3×3 matrix can now be created using this information.

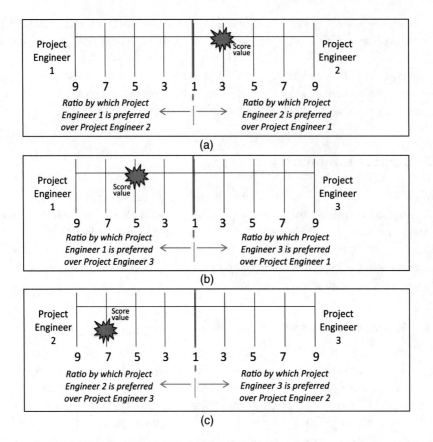

Fig. 10-2. Pairwise comparison. (a) Project Engineer 1 compared to Project Engineer 2, (b) Project Engineer 1 compared to Project Engineer 3, and (c) Project Engineer 2 compared to Project Engineer 3.

The partial matrix is assembled as in Fig. 10-3 by using the pairwise comparison. The first step is to note that all diagonal elements are always 1, because each one shows a criterion compared with itself. Next, referring to Fig. 10-2a, we place 1/3 in the matrix where Project Engineer 1 is compared to Project Engineer 2, implying that Project Engineer 2 is favored 3:1 over Project Engineer 1. The value of 5 is placed where Project Engineer 1 is compared to Project Engineer 3, indicating that Project Engineer 1 is preferred 5:1 over Project Engineer 3, and the value of 7 is placed where Project Engineer 2 is compared to Project Engineer 3.

	Project Engr 1	Project Engr 2	Project Engr 3
Project Engr 1	1	1/3	5
Project Engr 2		1	7
Project Engr 3			1

Fig. 10-3. Partial matrix of pairwise comparisons

	Project Engr 1	Project Engr 2	Project Engr 3
Project Engr 1	1	1/3	5
Project Engr 2	3	1	7
Project Engr 3	1/5	1/7	1
Sum	21/5	31/21	13

Fig. 10-4. Full matrix of pairwise comparisons

Next, we complete the matrix using the reciprocal of each initial input value. This is because the upper triangle of the matrix is the inverse of the lower triangle. Thus, because the relationship between Project Engineer 1 and Project Engineer 2 is 1:3, the reciprocal of 1:3 describes the relationship between Project Engineer 2 and Project Engineer 1. All the elements in the comparison matrix are positive. The row values are i, and the column values are j. Thus, $a_{ji} = (1/a_{ij})$. The full matrix is given in Fig. 10-4.

By looking at the comparison matrix, we can see that Project Engineer 1 is five times better than Project Engineer 3, and Project Engineer 2 is three times better than Project Engineer 1, in the best judgment of the project manager. Thus, we can say that Project Engineer 1 has more importance than Project Engineer 3, but Project Engineer 2 has more importance than both Project Engineer 1 and Project Engineer 3. However, the question remains whether these weights are reasonable, or whether they are off the charts, poorly weighted, random, or simply unreasonable. To determine the answer, an evaluation of the consistency of the weights is needed. To get there, we have to go through a number of intermediate mathematical steps.

10.9.2 Geometric Mean

First, Saaty (1980) recommended finding the geometric mean, which he equated to the normalized eigenvector of the largest eigenvalue from the full matrix, because the geometric mean is a good surrogate for the normalized eigenvector of the largest eigenvalue. To accomplish this, each column is summed vertically, as shown in the bottom row of Figure 10-4.

Next, each value within each column is divided by the sum of that column to normalize it. The normalization of each column is demonstrated in Fig. 10-5. Note that now each column adds up to 1.

	Project Engr 1	Project Engr 2	Project Engr 3
Project Engr 1	5/21	7/31	5/13
Project Engr 2	15/21	21/31	7/13
Project Engr 3	1/21	3/31	1/13
Sum	1	1	1

Fig. 10-5. Normalization of the matrix

$$\begin{bmatrix} W1 \\ W2 \\ W3 \end{bmatrix} = \frac{1}{3} \begin{bmatrix} 5/21 & + & 7/31 & + & 5/13 \\ 15/21 & + & 21/31 & + & 7/13 \\ 1/21 & + & 3/31 & + & 1/13 \end{bmatrix} = \begin{bmatrix} 0.2828 \\ 0.6434 \\ 0.0738 \end{bmatrix}$$

Principal eigenvector (Priority vector)

Fig. 10-6. Averaging across rows

The next step is to take the average across the rows (see Fig. 10-6). Saaty (1980) stated that the averaging across each row (W) will be equivalent to the normalized principal eigenvector of mathematical matrices.

The principal eigenvector is also called the priority vector. This priority vector now reflects the relative weights of each alternative. Project Engineer 1 is 28.28% preferable, Project Engineer 2 is 64.34% preferable, and Project Engineer 3 is 7.38% preferable.

10.9.3 Largest Eigenvalue

Every mathematical matrix has multiple eigenvalues, and some of them may be imaginary[3]. Saaty (1980) recommended that, when using AHP methodology, the decision maker calculate the largest eigenvalue (λ_{max}) to verify the consistency of the weights. To calculate this value, determine the product of the sum of each column (from Fig. 10-4) and the average of each row (from Fig. 10-6).

$$\lambda_{max} = \left[\frac{21}{5} \times (0.2828) \right] + \left[\frac{31}{21} \times (0.6434) \right] + [13 \times (0.0738)] = 3.097 \quad (10\text{-}3)$$

Using MATLAB to find λ_{max} resulted in a value of 3.065, which is close enough to the geometric mean approach used earlier that the geometric mean can indeed serve as a surrogate for the actual eigenvalue. This surrogate can be easily calculated at an office using a simple calculator, and may be used when it is difficult or tedious to find the real, mathematical eigenvalue, for which specialty software is sometimes needed.

10.9.4 Consistency Index

λ_{max} is now used to calculate the consistency index (CI) to verify the weights. Saaty (1980) recommended that the CI be calculated as in Eq. (10-4):

$$CI = \frac{\lambda_{max} - n}{n - 1} = \frac{3.097 - 3}{3 - 1} = 0.0485 \quad (10\text{-}4)$$

where λ_{max} is the maximum eigenvalue of the weighted matrix, and n is the order of the matrix. In this example, the order of the matrix is 3. Thus, the CI = 0.0485.

---

[3]Multiples of the square root of negative 1.

10.9.5 Random Index and Consistency Ratio

To determine whether the weights follow consistent ratios, Saaty (1980) recommended the use of a consistency ratio defined by the ratio of the consistency index (CI) to a random index (RI). Thus:

$$CR = \frac{CI}{RI} \tag{10-5}$$

Selecting a random index is not all that straightforward, and different agencies and authors have used different reckoning values. A brief listing of RI values by different authors for $n = 3$ is given in Table 10-7. Nevertheless, even using these given RI values is not always straightforward. For instance, with the Alonso-Lamata RI, a linear fit must be used[4], as on the online AHP calculator BPMSG (Goepel 2014; BPMSG 2016; Alonso and Lamata (2006)), whereas with the Oak Ridge National Laboratory (ORNL), the RI value can be used directly (as shown in Table 10-8).

The criterion for the consistency ratio can be as tight or loose as the user prefers. This is similar to the level of significance used in statistics for proving or disproving a hypothesis. The traditional acceptance criterion for CR, α, is generally taken as 10% (or 0.10) (Saaty 1980). Using a low value for α indicates that only a low variation in

Table 10-7. RI Values from Various Authors for $n = 3$ (adapted from Alonso and Lamata 2006)

Author	Oak Ridge	Aguaron et al.	Alonso-Lamata	Forman	Golden-Wang	Lane-Verdini	Noble	Tumala-Wan	Wharton
No. of matrices	100	100,000	100,000	—	1,000	2,500	500	—	500
Random index (RI)	0.382	0.525	0.524	0.5233	0.5799	0.52	0.49	0.500	0.58

Table 10-8. Random Index used by Oak Ridge National Laboratory for $n = 3$ for Various Scale Ranges

Scale	1–5	1–7	1–9	1–15	1–20	1–90
RI	0.190	0.254	0.382	0.194	0.12	0.72

---

[4]$RI = \frac{\bar{\lambda}_{max} - n}{n-1}$; $\bar{\lambda}_{max}(n) = 2.7699n - 4.3513$; $CR = \frac{\lambda_{max} - n}{\bar{\lambda}_{max} - n}$

weights will be tolerated, whereas using a high value for α indicates that there can be more laxity in the judgment. If there is a close fit between the weights—indicated by $CR < \alpha$—then no weight is outlandishly low or high compared to another.

Moreover, the maximum accepted eigenvalue, λ_{max}, will change for different values of α. In our case, λ_{max} is 3.097 if the geometric mean is used, or 3.065 if MATLAB is used. Per Alonso and Lamata (2006), the minimum α should be approximately 10% for λ_{max} of 3.097 and 5% for λ_{max} of 3.065.

Using the BPMSG calculator mentioned previously, we obtain a CR of 6.8%, although this calculator uses the Alonso-Lamata values for RI. However, using the generally accepted RI values from ORNL (shown in Table 10-8), we find that

$$CR = 0.0485/0.382 = 0.1269 \tag{10-6}$$

which is acceptable for an assumed α of 15%. Should our α have been 10%, the consistency ratio we obtained (12.69%) would indicate that the spread of weights was looser than our criteria and that the subjective judgment was unreasonable. Hence, we would have to redo the entire example by reassigning appropriate weights and bringing the CR to 10% or less in order to demonstrate that the weight assignment was reasonable. In this case, because we choose an α of 15%, the weights are acceptable.

10.9.6 Sensitivity Analysis

Sensitivity analysis is another optional step to see what effect a slight change in an input value in the matrix will have on the end result. Different alternatives and scenarios can be fabricated by going back to the beginning of the problem to change some weights. Different results could spark further discussion and deepen the analysis. If the rankings do not change when the weights are altered, the results are robust. If the rankings do change, the results are sensitive. Sensitivity analysis is an important tool for verifying that the result is ideal.

10.10 AHP Advantage and Critiques

The great advantage of using AHP is the simplicity behind the procedure, allowing simple arithmetical manipulations to substitute for advanced mathematical analysis and enabling decision makers to make both simple and complex decisions. A first-time user of the process can easily arrive at an optimal decision by following the techniques set forth in this chapter. After completing the AHP, the decision maker is able to analyze the sensitivity and consistency of the end result, which creates more confidence in the decision-making process.

Although widely used and accepted, AHP also has its critics (Triantaphyllou and Mann 1995; Perez et al. 2006) because of the way pairwise comparisons are used and alternatives are evaluated. Belton and Gear (1983) stated that when an alternative is introduced that is similar to an already-existing alternative, the ranking of the existing alternative may, in fact, be reversed, though it is difficult to imagine how this can happen. Thus, they proposed that each column within the AHP be divided by the maximum entry in that column, introducing the *revised AHP*. Saaty (1994) later accepted the revised AHP, calling it the *ideal mode AHP*.

Because alternative-ranking reversal occurs when similar alternatives exist, Saaty (1987) and Saaty and Vargas (1993) suggested analyzing each alternative to ensure that no near-copies are created. Saaty (1987)stated that if a decision maker ranks one alternative within 10% of another, then one of the alternatives has the potential to be eliminated. Dyer (1990) later criticized this recommendation, claiming that it wasn't really necessary because the weighting system intrinsically takes care of the relative rankings.

10.11 Conclusion

AHP provides a simple mathematical approach to solving complex multicriteria decision making (MCDM) problems, not only in engineering but also in everyday applications. Although the decision maker must use best judgments that are partly objective and partly subjective to arrive at the final decision, simple pairwise comparisons can lead to determining the best possible alternative to reach the ultimate goal.

In order to determine the ideal alternatives and criteria, both the deductive approach and the systems approach should be used. The decision maker should be aware that the true ideal alternative when making a decision is unknown, and AHP is not a panacea. Also, we should note that the implementation of AHP via eigenvalues and the consistency index is merely a method to aid in the decision-making process, and other methods (such as common and uncommon sense) should also be used to arrive at the final decision.

Following the simple AHP process shown in this chapter will aid in decision making, providing a mathematical approach and thus analytical reasoning to arrive at an answer. Although there are critiques of this method, it appears likely that, with the right criteria and alternatives, AHP will determine the ideal alternative for weighted criteria.

10.12 Exercises

- Derive your own realistic AHP model using a 3×3 matrix, and test it for consistency.
- Apply AHP in conjunction with Lawrence Miles's satisfaction factors (presented in Ch 9) for ranking alternatives.

References

Alonso, J. A., and Lamata, M. T. (2006). "Consistency in the analytic hierarchy process: A new approach." *Int. J. Uncertainty Fuzziness Knowledge Based Syst.*, **14**(4), 445–459.

Belton, V., and Gear, A. E. (1983). "On a short-coming of Saaty's method of analytic hierarchies." *Omega*, **11**(3), 228–230.

BPMSG (Business Performance Management Singapore). (2016). "BPMSG AHP priority calculator." <http://bpmsg.com/academic/ahp_calc.php> (Jul. 5, 2016).

Dyer, J. S. (1990). "Remarks on the analytic hierarchy process." *J. Manage. Sci.*, **36**(3), 249–258

Forman, E. H., and Selly, M. A. (2001). *Decision by objectives: How to convince others that you are right*, World Scientific, River Edge, NJ.

Goepel, K. D. (2014). "BPMSG's AHP online system: Rational decision making made easy." <http://bpmsg.com/academic/documents/BPMSG-AHP-OS-2014-05-14.pdf> (Jul. 5, 2016).

Ishizaka, A., and Nemery, P. (2013). *Multi-criteria decision analysis: Methods and software*, Wiley, Somerset, NJ.

Perez, J., Jimeno, J. L., and Mokotoff, E. (2006). "Another potential shortcoming of AHP." *Top*, **14**(1), 99–111.

Saaty, T. L. (1977). "A scaling method for priorities in hierarchical structures." *J. Math. Psychol.*, **15**(3), 234–281.

Saaty, T. L. (1980). *The analytic hierarchy process: Planning priority setting, resource allocation*, McGraw-Hill, New York.

Saaty, T. L. (1982). *Decision making for leaders: The analytical hierarchy process for decisions in a complex world*, Lifetime Learning, Belmont, CA.

Saaty, T. L. (1987). "Rank generation, preservation and reversal in the analytic hierarchy decision process." *Decis. Sci.*, **18**(2), 157–177.

Saaty, T. L. (1994). *Fundamentals of decision-making and priority theory with the AHP*, RWS, Pittsburgh, PA.

Saaty, T. L., and Vargas, L. G. (1993). "Experiments on rank preservation and reversal in relative measurement." *Math. Comput. Modell.*, **17**(4/5), 13–18.

Singh, A. (2005). "Analysis of behavioral parameters in partnering." K. Sullivan and D. T. Kashiwagi, eds., Construction Procurement: The Impact of Cultural Differences and Systems on Construction Performance, Proc., Int. Symp. of CIB W92, Las Vegas, NV, CIB, Rotterdam, Netherlands, 337–345.

Teknomo, K. (2006). "Analytic hierarchy process (AHP) tutorial." <http://people.revoledu.com/kardi/tutorial/AHP> (May 25, 2014).

Triantaphyllou, E., and Mann, S. H. (1995). "Using the analytic hierarchy process for decision making in engineering applications: Some challenges." *Int. J. Ind. Eng. Theory Appl. Pract.*, **2**(1), 35–44.

CHAPTER 11

Project Planning—The OOPS Game

11.1 Introduction

The OOPS game was developed to explore decision-making uncertainty and its implications for project management. The game is built on simple relationships and scoring rules; however, these relatively basic rules yield 64 different game strategies, resulting in 23,224,320 possible scores. The concepts, lessons, and strategies that can be learned from understanding the OOPS game reinforce common ideas of successful project management, which in general emphasizes the importance of planning and risk-taking.

11.2 Origins

Gregory Howell and Mike Vorster developed the OOPS games on a road trip from Blacksburg to Newport News, Virginia, around 1991. Howell wanted to invent a game that would put his personal experience into the context of a project. His "oops" experience happened when he was attempting to drive to an important appointment. When Howell got in the car to leave, he realized that he was low on fuel and not certain he had enough to make it to his appointment; however, if he took the time to stop for gas, he was not sure if he would be on time for the appointment (Howell, personal communication 2007). In such a situation, how does one decide what to do, and how does one gain experience in making such decisions? Howell's motivation was to develop a game demonstrating decision making under uncertainty in the context of a project. The relatively simple game that he developed requires few materials.

11.3 How to Play the OOPS Game

The goal of the OOPS game is to construct a project for the lowest cost (Howell, unpublished flyer 2006). The "Project" consists of nine activities that need to be

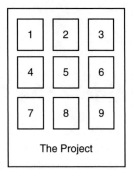

Fig. 11-1. Project card arrangement

arranged as shown in Fig. 11-1. Common playing cards can be used to represent the activities in the project (ace through 9). The project is constructed in nine draws.

First shuffle the cards and place the stack of cards face down. Remove the top card and place it face up in the Project area, in accordance with Fig. 11-1. The first card is always placed straightaway into the project area and always has a cost of $1. Subsequent cards can be placed in the project area only if the common-edge constraint is satisfied. For example, if the first card is a 2, there are only three possible successor activities: 1 (ace), 3, and 5. If the 2 and then the 5 are constructed, the possible successor activities are 1, 3, 4, 6, and 8.

Do not get distracted by the numbers on the cards, which serve only as an easy way to reference their location in Fig. 11-1. Once having become more familiar with the OOPS game, one will see that anything could be substituted for the numbers. What is important is how the activity relationships are defined in Fig. 11-1 or in a similar figure. Fig. 11-2 shows the layout and costs for each path of the OOPS game.

After the first draw, a decision needs to be made on the strategy for constructing the next activity. The options are either BUILD or PLAN. BUILD is the riskier of the two options; it is potentially either the cheapest strategy or the most expensive. If the player attempts to BUILD and is unsuccessful, the failure is called an OOPS for the mistake made (hence, the name of the game). The PLAN strategy always yields the same cost for constructing an activity. Fig. 11-2 defines the potential costs of each strategy; the number next to each path indicates the cost of that path. From a quick glance at Fig. 11-2, one can see that to PLAN always costs $2 (from Yard to Planning to Project), whereas a successful attempt to BUILD costs $1 (Yard to Project), and an unsuccessful BUILD attempt (a.k.a. an OOPS) costs $4 (Yard to OOPS and back to Project). Fig. 11-3 shows a flow chart for the OOPS gameplay.

The heart of the OOPS game is balancing the number of PLANs and BUILDs to construct the project for the least cost. The lowest cost for which the project can be constructed is $9; in the worst possible scenario, in which the player never PLANs and the risk of BUILD never pays off, the total bill can be $25, nearly triple the cost!

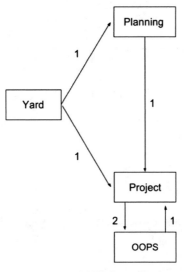

The OOPS Game Diagram

Fig. 11-2. OOPS game layout with scores

11.3.1 OOPS Game Example

Suppose the cards are shuffled and are in the sequence shown in Table 11-1. The first line in Table 11-1 shows the number of times a card is drawn, nine in total. The second line shows the card sequence; however, in reality the sequence would be unknown. The third line shows the score associated with each card. The fourth line shows the sequence in which the cards appear in the project space. The last line shows the total score for the game.

The first card—in this case the 1 card (ace)—is placed in the Project, at a cost of $1. Now the player must decide whether to PLAN or BUILD. There are eight decision points in the game at draws two through nine. In this case, the player chooses to BUILD. The next card is a 5. This card does not share a common edge with the 1, so it cannot be placed in the Project; therefore, it becomes an OOPS. The costs associated with a card that is OOPS-ed are $1 + $2 = $3. Therefore, the total score at this point is $4.

Now the player chooses to PLAN the next card. It is a 2, which can be placed in the Project. Because this activity was PLANned, it cost $1 + $1 = $2, making the total score at this point $6. Now that the 2 card is in place, the 5 card, which shares a common edge with 2, can be brought back from OOPS and placed in the Project; it costs $1 to bring a card back from OOPS to the Project. The running total cost rolls on to $7.

Now, at the fourth draw, the player decides to BUILD. The next card drawn is a 3. It can be placed in the Project, so the gamble of choosing to BUILD pays off, making the cost only $1; therefore, the total score now accumulates to $8. Again,

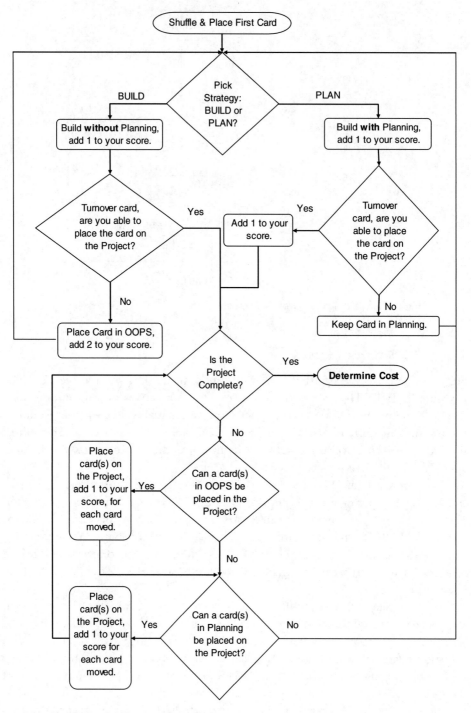

Fig. 11-3. OOPS game flow chart

Table 11-1. Example of an OOPS Game

Draw	1	2	3	4	5	6	7	8	9
Card Sequence	1	5	2	3	6	8	9	4	7
Score Incurred	1	4	2	1	1	1	1	1	1
Project Sequence	1	2	5	3	6	8	9	4	7
Hand Score	13								

at the fifth draw, the player chooses to BUILD. The next card is a 6, which can be placed in the Project at the cost of $1, making the total $9. The remaining four cards will be played by choosing BUILD. For this particular card sequence, the gamble pays off, and the total cost for each remaining card is $1, for a final score of $13.

11.4 OOPS Game Analysis

All the statistics quoted in this chapter were generated by playing the OOPS game in every possible way and recording the score. MATLAB was used to develop 71 programs and two user-defined functions (accounting for 167 pages of code) to play the game quickly and track the score for each attempt. In developing the programs, the following analytical strategies and properties of the OOPS game were considered.

11.4.1 Reducing the Scope of Analysis

First, during the game, a decision needs to be made only six times, not nine. This is because the first card is always placed directly in the Project area, and the player can always BUILD the last two cards without risk of OOPSing. When three cards remain, there is still a (small) risk of OOPSing (e.g., when 1, 2, and 4 remain, if one attempts to BUILD but draws the 1). Table 11-2 shows all the possible locations for two cards remaining (symmetry covers all examples not explicitly shown). Thus, there is no risk of OOPSing when two cards remain.

A less obvious situation is when two cards remain in the Yard, yet three spaces are left in the Project. The third card is either an OOPS or a PLAN. Table 11-3 shows one possible sequence of cards that illustrates this situation. After the sixth draw, the choice to BUILD is made. The seventh card is a 7, so it gets OOPS-ed. There are three remaining spaces in the Project and two remaining cards in the Yard. However, as the Project in Table 11-4 shows, the probability of success for the last two cards is still 100%.

This may appear to be a trivial reduction; however, it is hugely significant in terms of reducing the number of computations the computer needs to perform.

Table 11-2. Two-Card Combinations

1	2	3
4		
7	8	9

		3
4	5	6
7	8	9

	2	3
4		6
7	8	9

	2	3
4	5	6
7	8	

	2	
4	5	6
7	8	9

	2	3
4	5	6
7		9

Table 11-3. Example of a Sequence in Which Three Spaces Remain in the Project and Two Cards Remain in the Yard

Draw	1	2	3	4	5	6	7	8	9
Card Sequence	9	6	5	3	2	1	7	4	8
Score Incurred	1	1	1	1	1	1	4	1	1
Hand Score	12								

Table 11-4. Project for a Two-Card Example

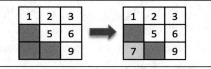

Clearly, a decision point exists each time a card is lifted for play: The decision is either to PLAN or to BUILD. If all the possible combinations of strategies for nine decision points ARE determined, the number of possible games that can be played is calculated using Eq. (11-1):

$$(\text{possible sequences of 9 cards}) \times (\text{possible strategies for 9 decision points})$$
$$= \text{total no. of games} \tag{11-1}$$

The number of possible sequences for nine cards is shown in Eq. (11-2):

$$9! = 362,880 \tag{11-2}$$

The number of possible strategies for nine decision points, where $n = 9$ represents the decision points and k represents the number of times PLAN is adopted, is calculated using Eqs. (11-3)–(11-5):

$$\sum_{k=0}^{n} C_k^n = \sum_{k=0}^{n} \binom{n}{k} = \sum_{k=0}^{n} \frac{n!}{k!(n-k)!} \tag{11-3}$$

$$= C_9^9 + C_8^9 + C_7^9 + C_6^9 + C_5^9 + C_4^9 + C_3^9 + C_2^9 + C_1^9 + C_0^9 \tag{11-4}$$

$$= 1 + 9 + 36 + 84 + 126 + 126 + 84 + 36 + 9 + 1 = 512 \tag{11-5}$$

Hence, the total number of possible games is shown in Eq. (11-6):

$$362,880 \times 512 = 185,794,560 \tag{11-6}$$

Therefore, the number of scores that need to be determined for a game with nine decision points is **185,794,560.**

Because it is possible to reduce the number of decision points to six, as explained previously, the number of scores that truly need to be determined is shown in Eq. (11-7):

$$\text{(possible sequences of 9 cards)} \times \text{(possible strategies for 6 decision points)}$$
$$= \text{total no. of games} \tag{11-7}$$

The number of possible strategies for six decision points is calculated using Eqs. (11-8) and (11-9):

$$= C_6^6 + C_5^6 + C_4^6 + C_3^6 + C_2^6 + C_1^6 + C_0^6 \tag{11-8}$$

$$= 1 + 6 + 15 + 20 + 15 + 6 + 1 = 64 \tag{11-9}$$

Hence, the total number of possible games is shown in Eq. (11-10):

$$362,880 \times 64 = 23,224,320 \tag{11-10}$$

Therefore, the number of scores that need to be determined for six decision points is **23,224,320 (23 million +).** This reduces the number by a factor of eight [compared to Eq. (11-6)], decreases the amount of programming needed, and significantly shortens the time needed to run the program.

11.4.2 Approach of the Analysis

The approach taken to determine the score for each possible strategy for each possible hand was first to determine the $9 \times 362{,}880$ matrix of scores for each card by playing the game in the riskiest possible way (selecting BUILD at each of decision points two through seven. Table 11-5 summarizes this matrix. Note that the matrix will be filled only with scores of $1 and $4, because one can only either successfully BUILD or OOPS each card.

Once this matrix is determined, it will be modified to accommodate the PLAN strategy at each decision point. Thus, the modification will consist of replacing the 1s and 4s with 2s. At the end of the analysis, 64 matrices will be filled with the scores for each card and each strategy, amounting to a total of $362{,}880 \times 64 = 23{,}224{,}320$ scores. (Table 11-6 visually summarizes all the strategies.) Each row of the matrix is summed, and the frequency of each score is shown on a histogram for the specific strategy. The final grouping, which appears later in Table 11-11, shows the data will also be grouped by the number of PLANs (i.e., $6\overline{6}$, $6\overline{5}$, $6\overline{4}$, $6\overline{3}$, $6\overline{2}$, $6\overline{1}$, $6\overline{0}$).

It is important for understanding the approach of the analysis to recognize that the score associated with each card is independent of when the card is placed in the Project area. The score associated with each card depends on two factors: which strategy is chosen (PLAN or BUILD) and where in the stack of cards it is located relative to its edge-constraint neighbors. For instance, if the first card placed in the Project area is a 1, and the next card drawn is a 9, the 9 will not be placed in the Project until at least the fifth draw. However, the score associated with the card becomes known during the second draw, because the score for the 9 will be either $2 or $4, depending on whether one chooses PLAN or BUILD at that time. (The score for the 9 in this sequence of cards cannot be $1 because none of its edge-constraint neighbors is ahead of it in the sequence.)

Table 11-5. All Scores Matrix

All_Score Matrix	Column									
Row	1	2	3	4	5	6	7	8	9	Score
1	1	1	1	1	1	1	1	1	1	9
2	1	1	1	1	1	1	1	1	1	9
3	1	1	1	1	1	1	1	1	1	9
4	1	1	1	1	1	1	1	1	1	9
⋮	⋮								⋮	⋮
⋮	⋮								⋮	⋮
⋮	⋮								⋮	⋮
362,877	1	4	4	4	4	4	1	1	1	24
362,878	1	4	4	4	4	4	1	1	1	24
362,879	1	4	4	4	4	4	1	1	1	24
362,880	1	4	4	4	4	4	1	1	1	24

Table 11-6. Summary of PLAN/BUILD Strategies

PLAN / BUILD	Sequence Code	Strategy #	Card Position 1 2 3 4 5 6 7 8 9	Number of Hands (N)	Mean	SD	Median	Mode	Low	High
Six Take Zero (6T0): All BUILD	1	1		362,880	15.822	4.024	15.000	15.000	9.000	27.000
Total for 6T0:				**362,880**	15.822	4.024	15.000	15.000	9.000	27.000
Six Take One (6T1): PLAN one	1	1.a	2	362,880	14.822	3.501	13.000	13.000	10.000	25.000
time over six decision points.	2	1.b	3	362,880	15.084	3.401	16.000	13.000	10.000	25.000
	3	1.c	4	362,880	15.410	3.352	16.000	16.000	10.000	25.000
	4	1.d	5	362,880	15.808	3.401	16.000	16.000	10.000	25.000
	5	1.e	6	362,880	16.260	3.590	16.000	16.000	10.000	25.000
	6	1.f	7	362,880	16.727	3.916	16.000	16.000	10.000	25.000
Total for 6T1:				**2,177,280**	15.685	3.593	16.000	16.000	10.000	25.000
Six Take Two (6T2): PLAN	1	2.a	8	362,880	14.080	2.874	14.000	14.000	11.000	23.000
twice over six decision points.	2	2.b	9	362,880	14.410	2.801	14.000	14.000	11.000	23.000
	3	2.c	10	362,880	14.671	2.715	14.000	14.000	11.000	23.000
	4	2.d	11	362,880	14.808	2.840	14.000	14.000	11.000	23.000
	5	2.e	12	362,880	15.260	3.037	14.000	14.000	11.000	23.000
	6	2.f	13	362,880	15.070	2.747	14.000	14.000	11.000	23.000
	7	2.g	14	362,880	15.727	3.388	14.000	14.000	11.000	23.000
	8	2.h	15	362,880	15.522	2.938	14.000	14.000	11.000	23.000
	9	2.i	16	362,880	15.395	2.727	14.000	17.000	11.000	23.000
	10	2.j	17	362,880	15.989	3.287	17.000	14.000	11.000	23.000
	11	2.k	18	362,880	15.848	2.902	17.000	17.000	11.000	23.000
	12	2.l	19	362,880	16.314	3.238	17.000	17.000	11.000	23.000
	13	2.m	20	362,880	16.246	2.983	17.000	17.000	11.000	23.000
	14	2.n	21	362,880	16.713	3.293	17.000	17.000	11.000	23.000
	15	2.o	22	362,880	17.165	3.494	17.000	17.000	11.000	23.000
Total for 6T2:				**5,443,200**	15.548	3.141	14.000	14.000	11.000	23.000
Six Take Three (6T3): PLAN	1	3.a	23	362,880	13.671	2.145	12.000	12.000	12.000	21.000
three times over six decision	2	3.b	24	362,880	14.070	2.159	15.000	12.000	12.000	21.000
points.	3	3.c	25	362,880	14.522	2.364	15.000	15.000	12.000	21.000
	4	3.d	26	362,880	14.395	2.115	15.000	15.000	12.000	21.000
	5	3.e	27	362,880	14.989	2.751	15.000	15.000	12.000	21.000
	6	3.f	28	362,880	14.848	2.302	15.000	15.000	12.000	21.000
	7	3.g	29	362,880	14.657	2.042	15.000	15.000	12.000	21.000
	8	3.h	30	362,880	15.314	2.678	15.000	15.000	12.000	21.000
	9	3.i	31	362,880	15.110	2.219	15.000	15.000	12.000	21.000
	10	3.j	32	362,880	15.246	2.379	15.000	15.000	12.000	21.000
	11	3.k	33	362,880	15.576	2.591	15.000	15.000	12.000	21.000
	12	3.l	34	362,880	15.713	2.723	15.000	15.000	12.000	21.000
	13	3.m	35	362,880	15.508	2.289	15.000	15.000	12.000	21.000
	14	3.n	36	362,880	16.165	2.937	15.000	15.000	12.000	21.000
	15	3.o	37	362,880	15.975	2.628	15.000	15.000	12.000	21.000
	16	3.p	38	362,880	15.833	2.292	15.000	18.000	12.000	21.000
	17	3.q	39	362,880	16.427	2.836	15.000	15.000	12.000	21.000
	18	3.r	40	362,880	16.300	2.611	15.000	18.000	12.000	21.000
	19	3.s	41	362,880	16.752	2.802	18.000	18.000	12.000	21.000
	20	3.t	42	362,880	17.151	2.890	18.000	18.000	12.000	21.000
Total for 6T3:				**7,257,600**	15.411	2.655	16.000	16.000	12.000	21.000
Six Take Four (6T4): PLAN four	1	4.a	43	362,880	13.657	1.351	13.000	13.000	13.000	19.000
times over six decision points.	2	4.b	44	362,880	14.110	1.556	13.000	13.000	13.000	19.000
	3	4.c	45	362,880	14.576	2.004	13.000	13.000	13.000	19.000
	4	4.d	46	362,880	14.508	1.620	13.000	13.000	13.000	19.000
	5	4.e	47	362,880	14.975	2.025	16.000	13.000	13.000	19.000
	6	4.f	48	362,880	14.833	1.599	16.000	16.000	13.000	19.000
	7	4.g	49	362,880	15.427	2.253	16.000	16.000	13.000	19.000
	8	4.h	50	362,880	15.300	1.982	16.000	16.000	13.000	19.000
	9	4.i	51	362,880	15.095	1.534	16.000	16.000	13.000	19.000
	10	4.j	52	362,880	15.752	2.192	16.000	16.000	13.000	19.000
	11	4.k	53	362,880	15.562	1.907	16.000	16.000	13.000	19.000
	12	4.l	54	362,880	16.014	2.108	16.000	16.000	13.000	19.000
	13	4.m	55	362,880	16.151	2.278	16.000	16.000	13.000	19.000
	14	4.n	56	362,880	16.413	2.187	16.000	16.000	13.000	19.000
	15	4.o	57	362,880	16.738	2.194	16.000	19.000	13.000	19.000
Total for 6T4:				**5,443,200**	15.279	2.134	16.000	16.000	13.000	19.000
Six Take Five (6T5): PLAN five	1	5.a	58	362,880	14.095	0.526	14.000	14.000	14.000	17.000
times over six decision points.	2	5.b	59	362,880	14.562	1.170	14.000	14.000	14.000	17.000
	3	5.c	60	362,880	15.014	1.419	14.000	14.000	14.000	17.000
	4	5.d	61	362,880	15.413	1.497	14.000	14.000	14.000	17.000
	5	5.e	62	362,880	15.738	1.481	17.000	17.000	14.000	17.000
	6	5.f	63	362,880	16.000	1.414	17.000	17.000	14.000	17.000
Total for 6T5:				**2,177,280**	15.137	1.455	14.000	14.000	14.000	17.000
Six Take Six (6T6): All PLAN	1	6.a	64	362,880	15.000	0.000	15.000	15.000	15.000	15.000
Total for 6T6:				**362,880**	15.000	0.000	15.000	15.000	15.000	15.000
Total for All Strategies:				**23,224,320**	15.412	2.697	15.000	15.000	9.000	27.000

To modify the scores for the strategy in which one PLANs three times (second, third, and seventh draws), the scores in columns 2, 3, and 7 are replaced with 2s; this is summarized in Table 11-7. For each strategy outlined in Table 11-6, the scores are modified in a similar fashion.

11.5 Results of Analysis

11.5.1 All Scores

Fig. 11-4 shows the histogram for every possible score for the 23,224,320 OOPS games, and Table 11-8 displays the data used to generate the histogram. The mean score for the OOPS games is $15.412 ($\sigma = 2.697$, $N = 23,224,320$; note that N in this case is the population size, not the number of decision points (n), as described in Section 11.4.1), with a low score of $9 and a high of $27. All integer scores in between are possible except for $26, which is impossible to attain. The median and mode scores are both $15.

Note that if one adopts the 6T6 method and PLANs at each draw, the score will be $15 ($\sigma = 0$, $N = 362,880$), as shown in Table 11-11. This strategy involves no risk, and the card sequence has no impact on the outcome. Achieving a score lower than $15 would be the result of successfully adopting a riskier strategy.

Table 11-7. Strategy # 26: Six Take Three Score Matrix

Six Take Three	Column									
Row	1	2	3	4	5	6	7	8	9	Score
1	1	2	2	1	1	1	2	1	1	12
2	1	2	2	1	1	1	2	1	1	12
3	1	2	2	1	1	1	2	1	1	12
4	1	2	2	1	1	1	2	1	1	12
⋮	⋮							⋮	⋮	⋮
⋮	⋮							⋮	⋮	⋮
⋮	⋮							⋮	⋮	⋮
362,877	1	2	2	4	4	4	2	1	1	21
362,878	1	2	2	4	4	4	2	1	1	21
362,879	1	2	2	4	4	4	2	1	1	21
362,880	1	2	2	4	4	4	2	1	1	21

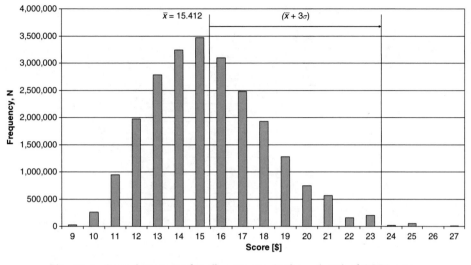

Fig. 11-4. Score histogram for all strategies (1 through 64) of OOPS game,
$N = 23,224,320$

11.5.2 All-BUILD Strategy

The result of the analysis if the decision to BUILD is made at each draw yields a mean score of $15.822 ($\sigma = 4.023$, $N = 362,880$), with a low score of $9 and a high of $27. Only integer scores of 9, 12, 15, 18, 21, 24, and 27 are possible, whereas it is impossible to score 10, 11, 13, 14, 16, 17, 19, 20, 22, 23, 25, or 26 with this strategy. The median and mode scores are $15. Fig. 11-5 is a histogram of the scores when the strategy of BUILD is used at each decision point, and Table 11-9 displays the data used for the histogram.

11.5.3 All PLAN Strategies

The result of the analysis for strategies that involve any PLANs (from 6T1 to 6T6) yields a mean score of $15.405 ($\sigma = 2.666$, $N = 22,861,440$), with a low score of $10 and a high of $25. All integer scores in between are possible except for $24. The median and mode scores are $15. Fig. 11-6 is a histogram of the scores for the sum of all the PLAN strategies, and Table 11-10 displays the data used for the histogram.

Similar histograms and tables were developed for all 64 strategies outlined in Table 11-6; they will not all be displayed here because of space constraints. However, the right-hand columns of Table 11-6 show the mean, standard deviation, median, mode, and high and low scores for each strategy.

Table 11-8. Table of All Scores for OOPS Game

All Scores for OOPS game			
Score	**Frequency**	**%**	**Cum. %**
9	29,568.000	0.13%	0.13%
10	256,768.000	1.11%	1.23%
11	946,464.000	4.08%	5.31%
12	1,972,864	8.49%	13.80%
13	2,778,272	11.96%	25.77%
14	3,246,720	13.98%	39.75%
15	3,465,248	14.92%	54.67%
16	3,098,432	13.34%	68.01%
17	2,477,568	10.67%	78.68%
18	1,931,648	8.32%	86.99%
19	1,277,152	5.50%	92.49%
20	746,112	3.21%	95.70%
21	563,488	2.43%	98.13%
22	160,512	0.69%	98.82%
23	203,616	0.88%	99.70%
24	14,784	0.06%	99.76%
25	49,344	0.21%	99.98%
26	0	0.00%	99.98%
27	5,760	0.02%	100.00%
Total Possible Hands	23,224,320		
Mean	15.412		
Median	15.000		
STD	2.697		

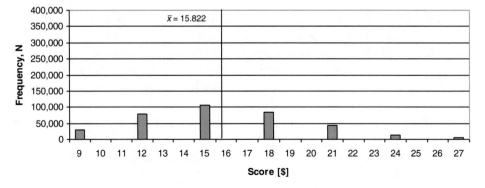

Fig. 11-5. Score histogram for BUILD on every draw (6T0), $N = 362,880$

Table 11-9. Score Breakdown for BUILD on Every Draw Strategy

All BUILD Scores			
Score	**Frequency**	**%**	**Cum. %**
9	29,568	8.15%	8.15%
12	79,360	21.87%	30.02%
15	106,144	29.25%	59.27%
18	83,968	23.14%	82.41%
21	43,296	11.93%	94.34%
24	14,784	4.07%	98.41%
27	5,760	1.59%	100.00%
Total Possible Hands	362,880		
Mean	15.822		
Median	15.000		
STD	4.023		

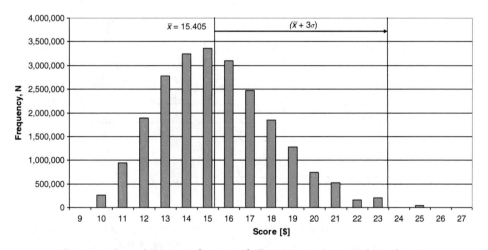

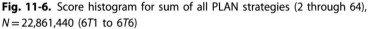

Fig. 11-6. Score histogram for sum of all PLAN strategies (2 through 64), $N = 22{,}861{,}440$ (6T1 to 6T6)

11.6 Analysis of Data

Comparing the all-BUILD strategy to the sum of all the PLAN strategies reveals a slight improvement in the mean score when any PLANing is done, as shown in Table 11-11. However, with each decision to PLAN, the minimum potential cost of the project increases. Conversely, with each decision to PLAN, the maximum potential cost of the project decreases. Thus, PLAN scores are bounded by $10 and $25, whereas BUILD scores vary from $9 to $27. Figs. 11-7 and 11-8 show the

Table 11-10. Summary Table for All PLAN Strategies

All PLAN Strategies (Summation)			
Score	**Frequency**	**%**	**Cum. %**
9	0	0.00%	0.00%
10	256,768	1.12%	1.12%
11	946,464	4.14%	5.26%
12	1,893,504	8.28%	13.55%
13	2,778,272	12.15%	25.70%
14	3,246,720	14.20%	39.90%
15	3,359,104	14.69%	54.59%
16	3,098,432	13.55%	68.15%
17	2,477,568	10.84%	78.98%
18	1,847,680	8.08%	87.07%
19	1,277,152	5.59%	92.65%
20	746,112	3.26%	95.92%
21	520,192	2.28%	98.19%
22	160,512	0.70%	98.89%
23	203,616	0.89%	99.78%
24	0	0.00%	99.78%
25	49,344	0.22%	100.00%
26	0	0.00%	100.00%
27	0	0.00%	100.00%
Total Possible Hands	22,861,440		
Mean	15.405		
Median	15.000		
STD	2.666		

cumulative score profiles for the all-BUILD strategy and for the sum of all the PLAN scenarios. The profiles make clear that only an all-BUILD strategy can yield the lowest possible score of $9, but, of course, this comes with the highest risk.

Figs. 11-9 and 11-10 show the mean scores, with range bars representing the standard deviation and high/low scores for all 23,224,320 games. As the number of decision points that are PLAN increases, the higher the lowest achievable score becomes (see Fig. 11-9). If one's approach to the game is risk averse, increasing the number of decision points he or she PLANs decreases the worst-case scenario cost for the project by lowering the highest possible project cost. Similarly, Fig. 11-10 shows that as the number of PLANs increases, the scores become more concentrated around the mean. This translates to more PLANning meaning greater certainty in the outcome. That said, PLANning does raise the minimum cost of the project.

Table 11-11. Summary of PLAN, BUILD, and Entire OOPS Game Strategies

| | Statistics | | | | | | |
	Mean	SD	Median	Mode	Low	High	N
All Scores	15.412	2.697	15.000	15.000	9.000	27.000	23,224,320
All Build, 6T0	15.822	4.024	15.000	15.000	9.000	27.000	362,880
6T1	15.685	3.593	16.000	16.000	10.000	25.000	2,177,280
6T2	15.548	3.141	14.000	14.000	11.000	23.000	5,443,200
6T3	15.411	2.655	15.000	15.000	12.000	21.000	7,257,600
6T4	15.279	2.134	16.000	16.000	13.000	19.000	5,443,200
6T5	15.137	1.455	14.000	14.000	15.000	17.000	2,177,280
6T6	15.000	0.000	15.000	15.000	15.000	15.000	362,880
Sum of PLANs	15.405	2.666	15.000	15.000	10.000	25.000	22,861,440

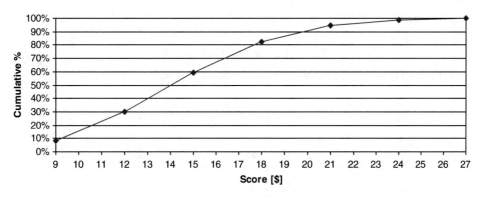

Fig. 11-7. Plot of cumulative percentage for all-BUILD strategy

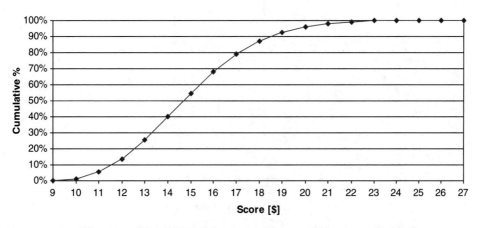

Fig. 11-8. Plot of cumulative percentage for sum of all PLAN strategies

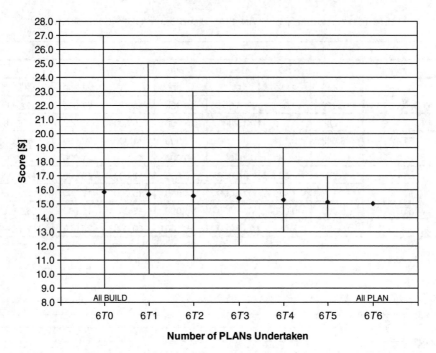

Fig. 11-9. Mean scores with high/low per number of times player PLANs

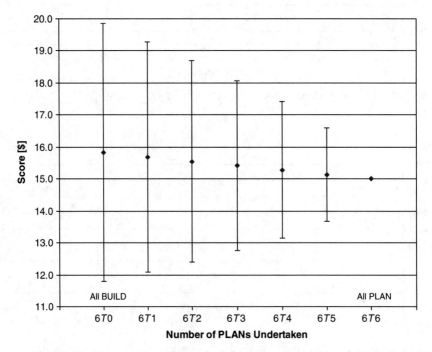

Fig. 11-10. Mean scores with standard deviation per number of times player PLANs

11.6.1 Lowest-Mean-Cost Strategy of 6T0 through 6T6 Strategies

The strategy that yields the lowest mean cost for the project is a $6T4$ (six take four) strategy in which the four PLANs occur at draws 2, 3, 4, and 5 (strategy 43, 4.a, Table 11-6). Table 11-12 summarizes the 4.a strategy. The mean score for the 4.a strategy is $13.657 (\sigma = 1.351, N = 362,880)$, with a low score of $13 and a high of $19 (the only other possible score is $16). The median and mode scores are $13.

Fig. 11-11 shows a histogram of the scores when the strategy of BUILD is used at each decision point, and Table 11-13 displays the data used for the histogram.

Table 11-12. Predetermined Strategy with Lowest Mean Score

| Plan | | Sequence | Sequence | Card Position | | | | | | | | | Statistics | | | |
Build		Code	Number	1	2	3	4	5	6	7	8	9	Mean	SD	Median	Mode
Six Take Four	1	4.a	43										13.657	1.351	13.000	13.000

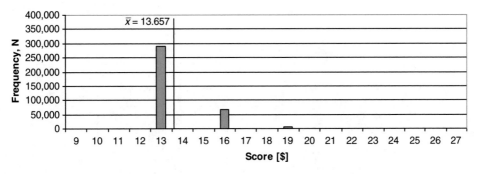

Fig. 11-11. Score histogram for strategy 43, 4.a, $N = 362,880$

Table 11-13. Summary Table for 4.a (Six Take Three)

Strategy 43, 4.a			
Score	**Frequency**	**%**	**Cum. %**
13	289,152	79.68%	79.68%
16	67,968	18.73%	98.41%
19	5,760	1.59%	100.00%
Total Possible Hands	362,880		
Mean	13.657		
Median	13.000		
STD	1.351		

11.6.2 BUILD When Probability of Success is Greater than 50% (P > 0.5)

The game can also be played with a different approach, in which the choices at the decision points are not predetermined. Instead, the player determines the chances of a successful BUILD attempt and only BUILDs if the chance of succeeding is greater than 50%; the player PLANs when the chance of succeeding is less than or equal to 50%. Consequently, a different set of scores results.

Table 11-14 shows the analysis of an example card sequence. Part A of Table 11-14 shows the example discussed in Section 11.3.1, with the addition of a row showing the probability of success; in that example, the player decided whether to PLAN or BUILD *arbitrarily*. Part B of Table 11-14 shows the example explained in the following paragraphs, in which the player uses the strategy of BUILD only when the probability of success is greater than 50%.

First, we will review Part A of Table 11-14: After the first card, which automatically goes into the Project, the probability of a successful BUILD on the second card needs to be determined. The numerator of this probability quotient is the number of locations in which one can successfully BUILD the next card: In this case, because the first card was a 1, the next card needs to be a 2 or a 4 in order to BUILD successfully, yielding a numerator of two. The denominator of the quotient is the

Table 11-14. Example with Probabilities

A

Draw	1	2	3	4	5	6	7	8	9
Card Sequence	1	5	2	3	6	8	9	4	7
Score Incurred	1	4	2	1	1	1	1	1	1
Project Sequence	1	2	5	3	6	8	9	4	7
Probability of Success	—	0.25	0.29	0.67	0.6	0.75	1.00	1.00	1.00
Ratio		1/4	2/7	2/3	3/5	3/4	3/3	2/2	1/1
Hand Score	13								

B

Draw	1	2	3	4	5	6	7	8	9
Card Sequence	1	5	2	3	6	8	9	4	7
Score Incurred	1	2	2	1	1	1	1	1	1
Project Sequence	1	2	5	3	6	8	9	4	7
Probability of Success	—	0.25	0.29	0.67	0.6	0.75	1.00	1.00	1.00
Ratio		1/4	2/7	2/3	3/5	3/4	3/3	2/2	1/1
Player Defined Probability	—	0.50	0.50	0.50	0.50	0.50	0.50	0.50	0.50
Hand Score	11								

number of remaining cards in the Yard. In this example, attempting to BUILD the second card has a 25% chance of succeeding (i.e., the numerator of the quotient is 2, and the denominator is 8). Sure enough, the next card was a 5, so the risk did not pay off.

Using the same logic, we can determine that the probability of success on the third card is $2/7 = 29\%$. The choice to PLAN helped mitigate the risk of OOPSing the next card. Although the PLAN did not pay off, the resulting punishment is not as severe as it would be for an OOPS. The probability of success on the next card (the fourth draw) is $4/6 = 67\%$. The huge improvement is because of the placement of the 5 in the Project area after the 2 was placed in the Project. The remaining probabilities are $3/5 = 60\%$, $3/4 = 75\%$, $3/3 = 100\%$, $2/2 = 100\%$, and $1/1 = 100\%$. Similar probabilities can be generated for any game and used to aid in the decision to PLAN or BUILD.

Revising Part A of Table 11-14 to have the player BUILD only when the chance of success is greater than 50% results in Part B of Table 11-14. After the 1 is placed, the chance of a successful BUILD is 25%, so the second card will be a PLAN ($25\% < 50\%$, therefore PLAN). This strategy pays off, because the card in the second draw is a 5, which is placed in Planning. The chance of a successful BUILD on the third draw is 29%, so the third card will be a PLAN as well. The 2 is placed in Planning, then into the Project area. Then the 5 is moved from Planning to the Project area. The chance of a successful BUILD on the fourth draw is 67%, so the player will be attempt to BUILD. The fourth card is a 3, so the BUILD attempt was successful. The chance of a successful BUILD on the fifth draw is 60%, so the strategy will be BUILD again. The fifth card is a 6, so the BUILD attempt succeeded. For the remaining draws, the chances of a successful BUILD are 75%, 100%, 100%, and 100%. For the example shown in Part B of Table 11-14, all the attempts to BUILD pay off, and the final score is $11.

If all 362,880 hands are played with this strategy (BUILD only when the chance of success is greater than 50%), the mean score is $13.418 ($\sigma = 2.014$, $N = 362{,}880$), with a low score of $10 and a high of $19 (all scores in between are possible), as summarized later in Table 11-17. The median score is $13, and the mode score is $12.

Fig. 11-12 shows a histogram of the scores for the greater-than-50% strategy, and Table 11-15 displays the data used for the histogram.

The scores are bounded between $10 and $19 (inclusive) in this scenario ($P > 0.5$). Thus, the range of the spread is less than that of either the all-BUILD strategy ($9 to $27; see Fig. 11-5) or the sum of all the PLAN strategies ($10 to $25; see Fig. 11-6). The standard deviation ($\sigma = 2.014$, $N = 362{,}880$) is also less than the sum of the PLAN strategies ($\sigma = 2.666$, $N = 22{,}861{,}440$) and the all-BUILD strategy ($\sigma = 4.023$, $N = 362{,}880$). Hence, this is an improvement in strategy compared to the sum of the PLAN strategies and the all-BUILD strategy. Similarly, the $P > 0.5$ strategy yields a lower mean and lower possible score than the 4.a strategy.

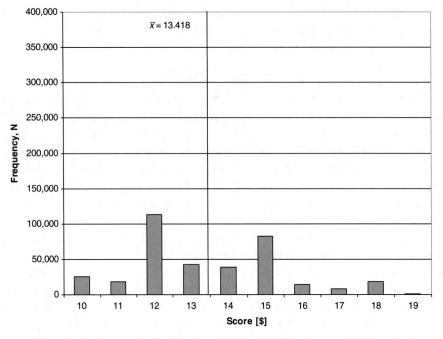

Fig. 11-12. Histogram of scores for choosing BUILD when probability of success is greater than 50%, $N = 362,880$

Table 11-15. Summary Table for Greater-than-50% Strategy

Build when the Probability of Success is Greater than 50%			
Score	**Frequency**	**%**	**Cum. %**
10	25,344	6.98%	6.98%
11	18,624	5.13%	12.12%
12	112,480	31.00%	43.11%
13	42,848	11.81%	54.92%
14	39,168	10.79%	65.71%
15	82,720	22.80%	88.51%
16	14,240	3.92%	92.43%
17	8,448	2.33%	94.76%
18	17,920	4.94%	99.70%
19	1,088	0.30%	100.00%
Total Possible Hands	362,880		
Mean	13.418		
Median	13.000		
STD	2.014		

11.6.3 BUILD When the Probability of Success is Greater than or Equal to 50% (P ≥ 0.5)

If the greater-than-50% strategy is modified to require a probability greater than *or equal to* 50% and the games are replayed, the scores generated are slightly different, although it is important to note that the change in strategy does not affect all hands. For instance, the example used for the greater-than-50% strategy is unaffected. However, the affected hands include those hands for which the lowest possible score (i.e., $9) is achievable. The mean score for the greater-than-or-equal-to-50% strategy is $13.775 ($\sigma = 2.528$, $N = 362{,}880$), with a low score of $9 and a high of $21 (all scores in between are possible), as summarized in Table 11-16. The median and mode scores are $14. Fig. 11-13 shows a histogram of the scores for the greater-than-or-equal-to-50% strategy, and Table 11-16 shows the data used for the histogram. Note that the mean score is higher than the greater-than-50% ($P > 0.5$) strategy; however, the lowest achievable score is the game minimum of $9.

The difference between the greater-than-50% and the greater-than-or-equal-to-50% strategies is that the latter is slightly riskier. The latter strategy has more decision points at which BUILD attempts are made. The addition of a few more opportunities to take a risk causes the mean score to increase; however, this strategy also allows the minimum score to decrease. By comparing the statistics for these two strategies, what can be observed is that no risk means no reward, and taking risks can improve scores.

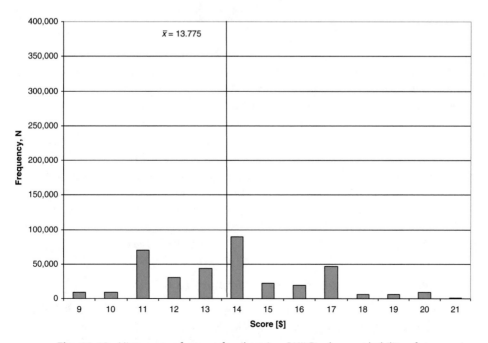

Fig. 11-13. Histogram of scores for choosing BUILD when probability of success is greater than or equal to 50%, $N = 362{,}880$

Table 11-16. Summary Table for Greater-than-or-Equal-to-50% Strategy

Build when the Probability of Success is Greater than or Equal to 50%			
Score	**Frequency**	**%**	**Cum. %**
9	9,408	2.59%	2.59%
10	9,408	2.59%	5.19%
11	70,528	19.44%	24.62%
12	30,208	8.32%	32.95%
13	43,488	11.98%	44.93%
14	89,440	24.65%	69.58%
15	21,888	6.03%	75.61%
16	19,104	5.26%	80.87%
17	46,816	12.90%	93.77%
18	6,528	1.80%	95.57%
19	5,760	1.59%	97.16%
20	9,216	2.54%	99.70%
21	1,088	0.30%	100.00%
Total Possible Hands	362,880		
Mean	13.775		
Median	14.000		
STD	2.528		

Learning when to PLAN and when to take risks can improve one's score, as demonstrated by the greater-than-50% strategy. When the player assesses at each draw the chances of a successful BUILD attempt, the mean score improves (from $13.657 to $13.418) over the best predetermined strategy, 4.a (see Table 11-12). In addition, although the highest possible score for both the greater-than-50% strategy and the 4.a strategy is $19, the greater-than-50% strategy has a minimum score of $10 (versus $13 for the 4.a strategy) yielding a lower possible score than the sum of all the PLAN scenarios. The mode for the greater-than-50% strategy is $12 (*frequency* = 112,480; 31% of the hands), whereas the mode for the 4.a strategy is $13 (*frequency* = 289,152; 80% of the hands). Thus, the most common score for the greater-than-50% strategy is lower than the most common score for strategy 4.a.

11.6.4 Varying the Threshold for Attempting to BUILD

If the OOPS game is played multiple times, varying the threshold probability for attempting to BUILD from 0% to 100% in increments of 25%, the mean score can be plotted against the threshold for risk, as shown in Fig. 11-14 (data in Table 11-17). A lower percentage translates to a higher-risk strategy; conversely, a higher percentage means a more conservative strategy. As a less risky strategy is adopted, the mean

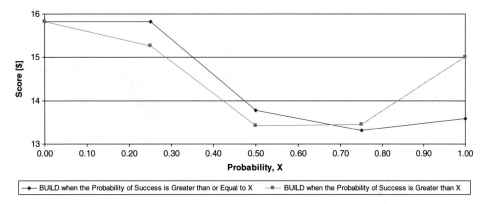

Fig. 11-14. Mean scores for OOPS game versus probability of BUILDing successfully

Table 11-17. Probability of Success for BUILD

BUILD when the Probability of Success is Greater than or Equal to X

	Statistics						
X	**Mean**	**SD**	**Median**	**Mode**	**Low**	**High**	**N**
0.00	15.822	4.024	15.000	15.000	9.000	27.000	362,880
0.25	15.822	4.024	15.000	15.000	9.000	27.000	362,880
0.50	13.775	2.528	13.000	12.000	9.000	21.000	362,880
0.75	13.316	1.408	13.000	13.000	11.000	16.000	362,880
1.00	13.591	0.893	14.000	14.000	11.000	15.000	362,880

BUILD when the Probability of Success is Greater than X

	Statistics						
X	**Mean**	**SD**	**Median**	**Mode**	**Low**	**High**	**N**
0.00	15.822	4.024	15.000	15.000	9.000	27.000	362,880
0.25	15.267	3.671	15.000	15.000	9.000	25.000	362,880
0.50	13.418	2.014	13.000	12.000	10.000	19.000	362,880
0.75	13.456	1.101	14.000	14.000	11.000	15.000	362,880
1.00	15.000	0.000	15.000	15.000	15.000	15.000	362,880

score decreases. However, choosing to PLAN improves the score only to a certain extent. At some point, the only way to further improve the project cost is to take risks. Somewhere between a 50% and a 75% strategy, less risk taken by the player actually increases the mean score; in other words, there is a point at which not taking risks is detrimental to the project cost! Also, a slightly more conservative approach than choosing BUILD when the probability of success is greater than 50%—BUILD when the probability of success is greater than or equal to 75%—yields an even lower mean score of \$13.316 ($\sigma = 1.408$, $N = 362,880$).

Table 11-18. Examples of Weighting Scheme

Plan		Sequence	Sequence	Card Position									Weighting
Build		Code	Number	1	2	3	4	5	6	7	8	9	Total
Six Take Two	13	2.m	20						1	2			3
Six Take Three	12	3.l	34			-2	-1		2				-1
Six Take Four	1	4.a	43		-3	-2	-1	1					-5
Six Take Six	1	6.a	64		-3	-2	-1	1	2	3			0

11.6.5 Choosing PLAN in the Beginning

Another strategy for improving the score is to adjust when one PLANs to the early draws. A weighting scheme is applied to each PLAN, within a planning strategy of −3 for the PLAN at the second draw to +3 for the PLAN at the seventh draw; see Table 11-18 for examples of applying weights.

With the weighting system in place, one can compare the mean scores for PLAN strategies based upon the total weighting. (Note that the weighting scheme is not sophisticated enough to compare strategies with different numbers of PLANs.) The lower the total weighting, the earlier the PLANs take place.

Fig. 11-15 shows the statistics for all Six Take Two strategies versus total weighting. The lower the total weighting is, the lower the mean score is for the strategy. By choosing to PLAN in the beginning, this results in a lower mean score. If

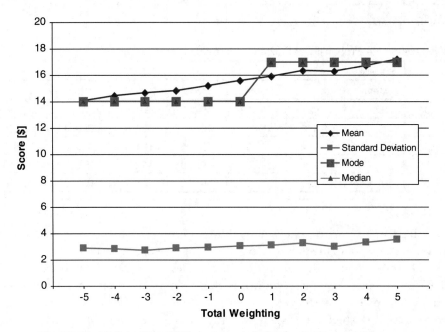

Fig. 11-15. Six Take Two (6T2) strategy statistics for PLAN weighting

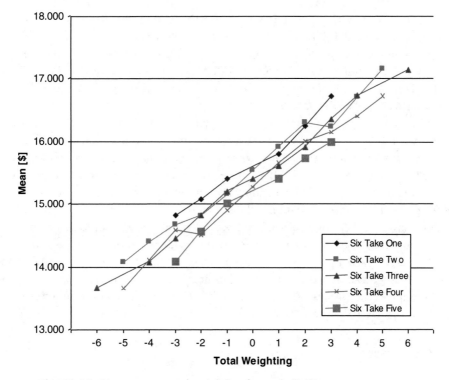

Fig. 11-16. Mean versus total weighting for each PLAN strategy

all the PLAN strategies (6T1 to 6T6, see Fig. 11-16) are compared, the trend is consistent throughout all the PLAN strategies. Thus, given that a certain number of PLANning decisions are to be made during a particular hand of the game, PLANning earlier yields a lower mean score than choosing PLAN later. Note that the lowest mean score from the 6T0 through 6T6 strategies was the PLAN sequence 4.a when the PLANs were at the first 4 draws. That PLAN sequence has the lowest total weight of all the 6T4 strategies, reinforcing that choosing PLAN early on lowers the mean score.

11.7 Conclusion

The most promising approaches to successfully playing the OOPS game are to PLAN at the early decision points and take calculated risks. Without taking any risks, the game's score cannot be lowered below $15. The strategy in which the player only attempts to BUILD when the chance of success is greater than or equal to 75% has the lowest mean score of any approach covered in this analysis; the worst score one can incur is $16 ($1 worse than choosing PLAN at every decision point approach; see Table 11-17). Furthermore, if the possibility of scoring $16 is unacceptable, changing the player's approach to BUILD only when the probability of success is greater than

75% results in a worst possible score of $15, with a mean score of $13.456 ($\sigma = 1.101$, $N = 362,880$). Thus, the greater-than-75% strategy is actually more conservative than the PLAN-at-every-draw strategy. This is a result of the all-PLAN strategy's blind commitment to choosing PLAN even when there is little risk of an OOPS. Thus, too much planning can be detrimental to the project cost, and risks are necessary to improve the score.

In reality, if project activities have relationships that require a particular sequence of work (note that a project can have more than one possible sequence), and attempting out-of-sequence work costs more than planning the work, the lessons learned from playing the OOPS game can decrease the total project cost (on average). It is not difficult to imagine ways to increase the complexity of the OOPS game, for example, changing the size of the Project from 3×3 cards to 5×5 cards or having a 3D Project area of $3 \times 3 \times 3$. The OOPS game could be extended to a full-blown construction project with enough consideration of activity relationships. As long as a particular work sequence needs to be followed and attempting out-of-sequence work costs more than planning the work, the average project cost can be improved by following strategies in line with those that result in lower mean scores in the OOPS game.

11.8 Exercises

Problem 1

Get a group of three people to play a card game with OOPS. One person should deal the cards, another should keep the score, and the third should be the referee. From a full deck of cards, select the ace through 9 of any suit. Find a large table with clear space on its surface and start to deal the cards from the Yard to the Project site. Leave enough space on the table for cards that go to Planning and those that go to OOPS. Record your score. Repeat the game a few more times. Try to strategize how to obtain the lowest score.

Problem 2

Divide the class into groups of three students. Let each group play the OOPS game 20 times. Record the scores. Then combine the resultant scores of all groups into a histogram. Find the standard deviation, mean, median, and mode. What do the results indicate?

Problem 3

Repeat Problem 2, but with each group using a different strategy. For example, one group can always PLAN, another group can always construct directly (BUILD), and other groups can try to PLAN at the beginning or at the end. Record the scores from all groups in a table. Perform a chi-square test to determine whether or not all the strategies have the same result.

CHAPTER 12

Tracking Performance through Control Charts

12.1 Introduction

This chapter studies control charts and their role in tracking performance. Control charts are relevant to risk analysis because of their ability—or lack thereof—to predict future performance. If control charts pertain directly to the consistency of a system or process, they can indicate the accuracy of predictions of future operating conditions. A system or process that is deemed out of control is inconsistent, which implies that there is some risk in the process. Predicted future performance, which is based on observed performance, will be relatively uncertain and undependable, and with greater uncertainty comes greater risk. Compare this to a system or process that is in control: Its process values are consistent, and future operations can be predicted with some accuracy, which reduces the amount of risk. Thus, control charts are both fundamental measures and indicators of risk.

12.2 The History of Control Charts

The control chart was developed in 1924 by Dr. Walter Shewhart ("Western Electric History," 2017; "Walter Shewhart," 2017; Smith 2011), who introduced the field of statistical process control (SPC), also referred to as statistical quality control (SQC). Shewhart worked as an inspector at Bell Telephone Laboratories (later AT&T), which was the Hawthorne plant of Western Electric Company (Best and Neuhauser 2006). His work with control charts helped improve Bell Telephone Laboratories' manufacturing process by identifying when adjustments were necessary.

Control charts gained wider recognition when W. Edward Deming expanded upon Shewhart's work and developed his own approach, Total Quality Management (TQM) (Deming 2000). One of the central principles of TQM is to "improve constantly and forever every process" (Walesh 2000). Control charts contribute to this objective by evaluating the consistency of a process.

12.3 Causes of Process Variation

A control chart is used to evaluate whether or not a process is statistically in control. The chart indicates the operating trend, emphasizing the process average and the variability of the process. Once the variability is known, its causes may be evaluated. The causes of process variations fall into two general categories, identified by Shewhart as *chance* (or random) and *assignable* (Best and Neuhauser 2006). Deming (2000) later renamed these *common* and *special* causes, respectively.

Chance (or random) causes, as the name indicates, occur simply by chance and cannot be assigned a particular reason. The variations they cause, which are generally found within the control limits, are built into the process. Chance causes may arise from factors such as the design of the process or the machinery used, over which the users may have little to no control. Such causes of variation can sometimes be mitigated (e.g., by choosing to use different tools), but they cannot be completely removed. *Common* is also an adequate label for these sources of variation because of their ubiquity. Chance causes exist in every process and thus are inevitable.

Assignable causes are so labeled because their sources can be identified. They are also referred to as *special* causes because they can be attributed to sources such as human error or freak occurrences. If the causes of variation are known (i.e., assignable), they can be addressed and eliminated until only chance or common causes remain. Because assignable causes are quickly removed by quality-conscious operators, the variations they cause typically fall outside the control limits and are not part of normal operations.

As an example, take the process of shopping for apples at the supermarket, for which the cost is being measured. Common causes of variation could result from slight differences in the number and weight of the apples being bought. In contrast, a special cause of variation might arise if the apples go on sale.

As previously noted, special causes of variation identified using control charts should be found and resolved; in other words, the process can and should be adjusted to improve the consistency and quality of the output.

Control charts can also prevent unnecessary adjustments. If a control chart was not used but some variation was observed, it might be assumed that something was wrong with the process and needed to be fixed. However, there is variation resulting from common causes in every process; in the absence of a control chart, such variation could be mistaken for variation resulting from special causes. If the observed variation arose solely from common causes, then the process was in control, and it would be unhelpful, and possibly detrimental, to try to fix a process that was not broken. Adjusting the process needlessly could harm its quality or even throw it out of control. Therefore, control charts can be used to deter unwarranted modifications, in addition to indicating when adjustments should be made.

12.4 Run and Control Charts, Upper Control Limit and Lower Control Limit, and Six Sigma

A *run chart*, also referred to as a time plot or trend chart, is a graph that displays data over time (i.e., time is on the *x*-axis). When a run chart takes standard deviation (σ) into account, such as with 3σ or 6σ, it becomes a control chart.

The *Six Sigma* method (Akpolat 2004) models its control charts on the normal, or Gaussian, distribution. In such a distribution, it is known that 99.73% of data falls within three standard deviations (3σ) of the mean. This means that locating control limits 3σ above and below the process average—for a total span of 6σ—results in only a 0.27% probability that a point will lie outside the control limits (Ehrlich 2002). As previously noted, common causes are typically located within the control limits. Any point that falls in the improbable locations outside the control limits is considered a variation resulting from special causes.

Without Six Sigma, a run chart can only show the pattern of the data; it cannot prove its consistency. This is the advantage of a control chart over a run chart. Specifically, a *control chart* is a run chart that defines and indicates limits known as the *upper control limit (UCL)* and *lower control limit (LCL)*. These limits are statistically determined according to the normal distribution, binomial distribution, or Poisson distribution, and placed 3σ above and below the process average. Additional control limits of 1σ and 2σ above and below the average can be created.

The UCL and LCL represent the actual operating characteristics of the system, not the desired specifications. However, these control limits can be useful when compared to limits that do represent the specifications—that is, the upper specification limit (USL) and lower specification limit (LSL) in process capability—because they show whether the actual operations are conforming to set standards.

12.5 Control Chart Concepts

Typically, samples of data are collected from a process running according to standard procedures (i.e., without tweaking) (Brassard 1988). The type of data available will determine which types of control chart are most appropriate. If the data samples are quantifiable (i.e., measurable), then *variable* control charts, such as the $\overline{X}$ chart and the R chart, are appropriate. The $\overline{X}$ chart uses averages, and the R chart uses range. If the data are qualitative—for example, a unit either does or does not meet standards—*attribute* control charts are more useful. This category of control chart is broken down further to cater to specific types of data (e.g., proportion or number defective, or number of nonconformities). One way or another, the information collected has to be transformed into numerical data.

The values from the data can be used to calculate and, subsequently, plot the process average and control limits, in addition to the process values. Because control charts use statistics, the accuracy and usefulness of these charts increases with the number of samples. Taking 20 to 25 groups of samples with about four or five measurements each (i.e., approximately 100 data points) is recommended for representative or predictive control limit calculations. Although different equations are used to calculate the UCL and LCL if a chart other than a Gaussian distribution is applicable, the original standard for control limits is based on the concept of Six Sigma. These other equations, which will be discussed subsequently, provide various ways of evaluating a distance of 3σ on each side of the average, making it a total span of 6σ.

Once the control chart, complete with its main elements—process values, process average, UCL, and LCL—has been established, the relationship between the process values and the process average and limits can be used to evaluate the state of control. If process values exceed the control limits or follow unrealistic patterns, the process may be categorized as *out of control.*

Control charts that determine the consistency of a system or process can indicate the accuracy of predictions of future operating conditions. A system or process that is deemed out of control is inconsistent: Its predicted future performance, which is based on the observed performance, is relatively uncertain and undependable, increasing risk. Compare this to a system or process that is in control: Its process values are consistent, and therefore its future operations can be predicted with relative accuracy, reducing risk.

As an example, we will evaluate risk in terms of vehicular safety. Let's assume a hypothetical Car A that periodically has a problem of sudden, unintended acceleration. Suppose that a control chart analysis of Car A was performed, for which the speed of the car was measured when the driver intended to travel at the speed limit—for example, 35 mph. When the control limits were calculated and the control chart graphed, most of Car A's process values were within the control limits, but a few were outside. Car A was determined to be out of control.

Now suppose a Car B that does not experience such acceleration problems or special problems of any other type. In a similar control chart analysis done for Car B, it was found to be in control. By looking at the chart, one could reasonably predict that Car B will continue to operate in control (i.e., move at the speed that the driver intended). The same cannot be said for Car A. At any time and without warning, Car A could accelerate beyond the driver's control. It might revert back to normal without causing any harm, but it might also send the car on a collision course. Although the result is unknown, the potential damage and injury represents a seemingly untenable risk to one's safety; Car A is, quantifiably, the riskier car.

Control charts can thus help individuals evaluate and track risk in processes or systems. Furthermore, because the charts can be adjusted—tightened—to improve quality, they can also help reduce the amount of risk.

Control charts are effectively applicable and heavily used in engineering—especially in manufacturing—because of their relevance to risk analysis and TQM. In any business, control charts can help achieve the goals of TQM by improving the quality of work done, which may in turn improve revenue.

12.6 Variable Control Charts

Variable control charts are used when samples are measured quantitatively (i.e., output is continuous), such as in terms of length, weight, or volume.

The two main types of variable control charts are the $\overline{X}$ chart and the R chart. For these control charts, measurement samples are collected over numerous time periods (e.g., days, weeks, or months). The X values represent the actual data collected. The average of the measurements over a given time period is expressed by the variable $\overline{X}$, which can be calculated using Eq. (12-1).

$$\overline{X} = \frac{X_1 + X_2 + \cdots + X_n}{n} \tag{12-1}$$

in which $X_1, X_2, \ldots, X_n$ = measurements taken in the first, second, $\ldots$, and nth samples, respectively; and n = number of samples.

The range of measurements within a specific time period is denoted by the variable R, which is the difference between the maximum and minimum X values, as indicated by Eq. (12-2).

$$R = X_{\max} - X_{\min} \tag{12-2}$$

The variable $\overline{R}$ is used to represent the average of all R values at any time of testing, provided that the number of samples taken, n, is greater than one.

Once $\overline{X}$ and R have been determined for all time periods, charts may be plotted. Both the $\overline{X}$ and R charts will have time on the horizontal axis and measurement on the vertical axis. Fig. 12-1 shows the advantage of using $\overline{X}$ and R charts instead of charts detailing all data collected. In Fig. 12-1(a), which shows all the measured samples plotted against time, the chart is complicated and difficult to evaluate. In Figs. 12-1(b and c), the $\overline{X}$ and R charts simplify the measured data into a more understandable format, clearly outlining the operational trend. Hence, run charts and control charts help to reduce chaos and narrow uncertainty. This allows the handling of risk in an organized manner.

Figs. 12-1(b and c) include several labeled lines, including $\overline{\overline{X}}$ and $\overline{R}$. $\overline{\overline{X}}$ represents the process average, or the average of the averages, and $\overline{R}$ is the average range. The values of these variables can be calculated using Eqs. (12-3) and (12-4).

$$\overline{\overline{X}} = \frac{\overline{X_1} + \overline{X_2} + \cdots + \overline{X_k}}{k} \tag{12-3}$$

$$\overline{R} = \frac{R_1 + R_2 + \cdots + R_k}{k} \tag{12-4}$$

in which k = number of subgroups (20–25 groups) taken over time.

After determining the process averages, $\overline{\overline{X}}$ and $\overline{R}$, the control limits can be computed and plotted on the control charts. Through statistical analysis, industry

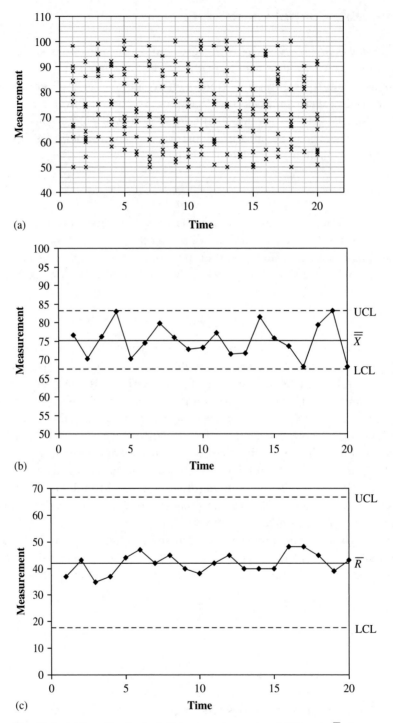

Fig. 12-1. (a) Sample chart of all measured samples; (b) Sample $\overline{X}$ chart; (c) Sample R chart

standards for plotting the limits, derived from the assumed probability distribution of the data, have been developed. This text will not dwell on the statistical derivations but will show practical applications of the charts across industry worldwide. For the $\overline{X}$ chart, Eqs. (12-5) and (12-6) are used to calculate the upper and lower control limits, respectively.

$$\text{UCL}_{\overline{X}} = \overline{\overline{X}} + \text{A}_2\overline{R} \tag{12-5}$$

$$\text{LCL}_{\overline{X}} = \overline{\overline{X}} - \text{A}_2\overline{R} \tag{12-6}$$

where A_2 is a factor for the $\overline{X}$ chart, as explained following Eq. (12-8).

In the charts discussed previously, which used a normal distribution, the upper and lower control limits were equally distant from the mean. For the R chart, however, the upper and lower control limits use different factors, as shown in Eqs. (12-7) and (12-8).

$$\text{UCL}_R = \text{D}_4\overline{R} \tag{12-7}$$

where D_4 is a factor for the upper limit of the R chart.

$$\text{LCL}_R = \text{D}_3\overline{R} \tag{12-8}$$

where D_3 is a factor for the lower limit of the R chart.

The factors for the $\overline{X}$ and R charts depend on the number of samples, n. Values for A_2, D_3, and D_4 may be found in Table 12-1.

Example 1: Rice Production

The Rice Company would like to evaluate its operating conditions for producing bags of rice. For eight weeks, the company weighed 10 bags of rice per week. The data collected are shown in Table 12-2a. Plot the variable control charts for the Rice Company's data.

Solution

Start by determining the average weight for Week 1. The weights given in Table 12-2a are the values of X. Because there were 10 sample measurements taken, $n = 10$. Thus, from Eq. (12-1), $\overline{X}_1 = \frac{4.25+3.50+3.50+4.50+4.00+4.50+4.75+4.25+3.25+3.50}{10} = 4.00$.

Next, determine the range for Week 1. X_{max} is 4.75, and X_{min} is 3.25. Thus, from Eq. (12-2), $R_1 = 4.75 - 3.25 = 1.50$, which is the range for the first week's data.

Now find the averages and ranges for each of the 10 weeks for which measurements were given. Table 12-2a can then be supplemented with these values; the results are shown in Table 12-2b.

At this stage, the average of the averages ($\overline{\overline{X}}$) and average range ($\overline{R}$) can be calculated.

Table 12-1. Factors for $\overline{X}$ and R Charts

Number of observations in subgroup (n)	Factors for X Chart	Factors for R Chart	
	A_2	Lower D_3	Upper D_4
2	1.880	0	3.268
3	1.023	0	2.574
4	0.729	0	2.282
5	0.577	0	2.114
6	0.483	0	2.004
7	0.419	0.076	1.924
8	0.373	0.136	1.864
9	0.337	0.184	1.816
10	0.308	0.223	1.777
11	0.285	0.256	1.744
12	0.266	0.283	1.717
13	0.249	0.307	1.693
14	0.235	0.328	1.672
15	0.223	0.347	1.653
16	0.212	0.363	1.637
17	0.203	0.378	1.622
18	0.194	0.391	1.609
19	0.187	0.404	1.596
20	0.180	0.415	1.585
21	0.173	0.425	1.575
22	0.167	0.435	1.565
23	0.162	0.443	1.557
24	0.157	0.452	1.548
25	0.153	0.459	1.541

There are $k = 8$ subgroups (i.e., weeks). Hence, from Eqs. (12-3) and (12-4), $\overline{\overline{X}} = \frac{4.00+4.10+3.75+3.85+3.90+4.15+3.80+4.35}{8} = 3.99$, and $\overline{R} = \frac{1.50+2.00+1.25+2.50+2.25+1.75+1.00+2.25}{8} = 1.81$. Now look up A_2, D_3, and D_4 for $n = 10$ in Table 12-1:

$$A_2 = 0.308$$

$$D_3 = 0.223$$

$$D_4 = 1.777$$

Now calculate the upper control limit (UCL) and lower control limit (LCL) for the average weight. From Eqs. (12-5) and (12-6), $\text{UCL}_{\overline{X}} = 3.99 + 0.308 \times (1.81) = 4.54$, and $\text{LCL}_{\overline{X}} = 3.99 - 0.308 \times (1.81) = 3.43$.

Table 12-2a. Weekly Weight Measurements for the Rice Company

	Time (weeks)							
	1	2	3	4	5	6	7	8
Bag #	Weight (lb)							
1	4.25	5.00	4.50	4.75	4.00	4.00	3.75	4.00
2	3.50	3.75	4.00	5.00	2.75	5.00	3.25	5.50
3	3.50	4.50	3.25	4.00	3.50	3.50	3.50	4.75
4	4.50	4.75	3.50	4.00	4.25	4.75	4.25	3.50
5	4.00	4.25	4.25	4.50	5.00	4.00	4.25	4.25
6	4.50	3.00	3.25	3.75	4.00	4.25	4.00	4.00
7	4.75	3.00	3.25	2.75	4.50	5.00	3.75	3.25
8	4.25	4.25	3.75	2.50	3.25	3.25	3.50	4.50
9	3.25	4.00	3.50	4.00	3.50	4.00	4.25	5.00
10	3.50	4.50	4.25	3.25	4.25	3.75	3.50	4.75

Table 12-2b. Weekly Averages and Ranges of Weight Measurements for the Rice Company

	Time (weeks)							
	1	2	3	4	5	6	7	8
	Weight (lb)							
$\overline{X}$	4.00	4.10	3.75	3.85	3.90	4.15	3.80	4.35
R	1.50	2.00	1.25	2.50	2.25	1.75	1.00	2.25

$\overline{\overline{X}} = 3.99$; $\overline{R} = 1.81$.

Find the UCL and LCL for the range using Eqs. (12-7) and (12-8): $UCL_R = 1.77 \times (1.81) = 3.22$, and $LCL_R = 0.233 \times (1.81) = 0.404$.

Using the calculated UCL and LCL, plot the $\overline{X}$ and R control charts (Figs. 12-2 and 12-3, respectively). The two charts show that the process is in control.

Although the combination of the $\overline{X}$ and R charts is the most widely-used set of variable control charts, there are alternative combinations, such as the $\overline{X}$ and S charts or the X and mR charts. For the $\overline{X}$ and S charts, the $\overline{X}$ chart remains the same, and the S chart is similar to the R chart but measures standard deviation instead of range. This combination of charts is recommended when the number of samples within a subgroup is greater than 10 (Ehrlich 2002).

The X and mR charts, also called individual and moving-range charts, consider individual data points. The X and mR charts are utilized when it is not reasonable to use rational subgroups (Ehrlich 2002), such as when the process would take too much time to produce enough samples to subgroup. Thus, the X chart is basically the $\overline{X}$ chart when $n = 1$, and the mR chart shows the moving range of successive data (i.e., the difference between one point and the next).

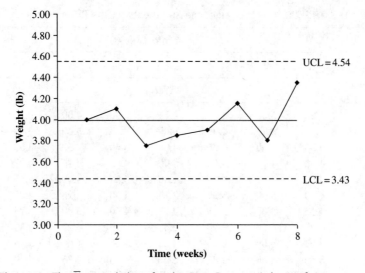

Fig. 12-2. The $\overline{X}$ control chart for the Rice Company's bags of rice

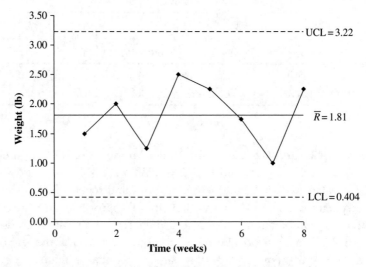

Fig. 12-3. The R control chart for the Rice Company's bags of rice

12.7 Attribute Control Charts

Attribute control charts are used when samples are assessed qualitatively (i.e., the output is discrete), such as defective or not defective, pass or fail, or good or bad. The four basic attribute control charts are the p chart, the np chart, the c chart, and the u chart.

- The *p* chart measures the *proportion* of defective units. Therefore, it does not require that all samples taken have the same sample size.
- The *np* chart measures the *number* of defective units. This type of chart works best with a constant sample size.
- The *c* chart measures the *number of nonconformities* when all samples taken have the same sample size.
- The *u* chart measures the *rate of nonconformities* or *number of nonconformities* with varying sample sizes.

Both the *p* chart and the *np* chart measure the same type of data—qualitative measures thought of in terms of all or nothing (Ehrlich 2002). A unit is either defective or not defective; it either meets or fails to meet requirements. The control limits for these charts are modeled upon the binomial distribution and are thus more effective for situations in which defective products are not rare (Xie et al. 2002).

Like the *p* and *np* charts, both the *c* chart and the *u* chart measure the same type of data, in this case the number of nonconformities. The number of non-conformities is a bit broader in concept than the defective units measured in the *p* and *np* charts. Whereas all defective units are nonconforming, not all noncon-formities necessarily render a unit defective: Small manufacturing errors might cause nonconformities without making the product as a whole defective. The control limits for these charts are based on the Poisson distribution, which is more commonly used for rare events (Xie et al. 2002) whose probability of occurrence is small.

Despite using different distributions, the *p* and *u* charts are similar in that they are appropriate for inconsistent sample sizes. Likewise, the *np* and *c* charts are comparable in that they work best for constant sample sizes.

12.7.1 The p Chart

The *p* control chart is used for occurrences that are not rare; it shows the proportion of a sample that is defective. Accordingly, the *p* chart shows time on the *x*-axis and percentage defective on the *y*-axis. Recall that for the $\overline{X}$ chart, individual measure-ments (X) within a subgroup were evaluated first. The same can be done for the *p* chart. The variable *p* for each subgroup represents the proportion defective within that subgroup. Mathematically, $p = \frac{number\ of\ rejects\ in\ subgroup}{number\ inspected\ in\ subgroup}$. Looking at the data as a whole, the proportion defective for all subgroups is $\overline{p} = \frac{total\ number\ of\ rejects}{total\ number\ inspected}$.

The UCL and LCL for the *p* chart, indicating variation from $+3\sigma$ to -3σ, can be calculated using Eqs. (12-9) and (12-10), which are derived from the binomial distribution.

$$\text{UCL}_p = \overline{p} + \frac{3\sqrt{\overline{p}(1-\overline{p})}}{\sqrt{n}} \qquad (12\text{-}9)$$

$$\text{LCL}_p = \bar{\bar{p}} - \frac{3\sqrt{\bar{\bar{p}}(1 - \bar{\bar{p}})}}{\sqrt{n}} \tag{12-10}$$

The variable n represents the number of samples, just as it did in the variable control charts. This can create some confusion, as there may be a multitude of sample sizes within one analysis, resulting in varying control limits. The general rule is to use the average sample size, $\bar{n}$, when the sample sizes all fall within ±20% of the average sample size. When a sample size varies by more than ±20% from the average, specific UCL and LCL values must be calculated for that subgroup (Brassard 1988). Both methods are utilized in the following examples.

Example 2a: Pen Power I

A manufacturer of ballpoint pens, Pen Power, wishes to improve its production. The company decides to make a control chart analyzing the proportion of defective ballpoint pens it produces. The daily count of defective pens and the total number inspected were collected, and the data are presented in columns 2 and 3 of Table 12-3a. From here, the specific proportion defective for each day is found as the ratio of column 2 to column 3 and presented in column 4 of Table 12-3a. The p values will be the process values plotted in the eventual control chart.

The proportion of all pens that were defective is represented by $\bar{\bar{p}}$ and can be calculated based on the total number of rejects and the total number inspected: $\bar{\bar{p}} = \frac{322}{2,500} = 0.129$.

However, finding the upper and lower control limits will not be as simple. Consider the 25 individual sample sizes shown. The set of data has an average sample size of 100 pens—2,500 total pens inspected, divided by 25 subgroups—but the individual sample sizes vary from 63 to 144. In fact, if the standard deviation of the sample size were to be calculated, it would be approximately 24.7. The general rule for the p chart says to evaluate individual limits for sample sizes falling outside of ±20% of the average sample size (100 pens). This translates to calculating specific UCL and LCL values for sample sizes that are not between 80 and 120 pens. When looking at the previous tables, notice that for 19 out of the 25 days when samples were collected, the sample size was either less than 80 pens or greater than 120 pens. Hence, one may use Eqs. (12-9) and (12-10) for the sample calculations below and then refer to Table 12-3b for the specific UCL and LCL for each day.

Day 1:

$$\text{UCL}_p = 0.129 + \frac{3\sqrt{0.129(1 - 0.129)}}{\sqrt{144}} = 0.213$$

$$\text{LCL}_p = 0.129 - \frac{3\sqrt{0.129(1 - 0.129)}}{\sqrt{144}} = 0.045$$

Table 12-3a. Daily Counts and Proportions of Defective Pens for Pen Power

Day (1)	Number of defective pens (2)	Number of inspected pens (3)	$P\,(4) = (2)/(3)$
1	16	144	0.111
2	24	128	0.188
3	12	122	0.098
4	10	107	0.093
5	14	98	0.143
6	16	125	0.128
7	7	78	0.090
8	10	122	0.082
9	13	78	0.167
10	19	130	0.146
11	17	76	0.224
12	21	127	0.165
13	8	63	0.127
14	12	92	0.130
15	16	131	0.122
16	12	121	0.099
17	11	79	0.139
18	12	95	0.126
19	7	69	0.101
20	13	76	0.171
21	12	75	0.160
22	6	115	0.052
23	11	101	0.109
24	9	69	0.130
25	14	79	0.177
Total	322	2500	

The upper and lower control limits will be vastly different and more extreme for Day 13 because the sample size changes drastically (from 144 pens on Day 1 to 63 pens on Day 13).

Day 13:

$$\text{UCL}_p = 0.129 + \frac{3\sqrt{0.129(1 - 0.129)}}{\sqrt{63}} = 0.255$$

$$\text{LCL}_p = 0.129 - \frac{3\sqrt{0.129(1 - 0.129)}}{\sqrt{63}} = 0.002$$

Similarly, each day's UCL and LCL are calculated based on the sample size taken that day. The p control chart is graphed using the data in Table 12-3b and presented in Fig. 12-4 for a visual analysis.

Table 12-3b. Upper and Lower Control Limits for Pen Power's p Control Chart

Day	UCL_p	LCL_p
1	0.213	0.045
2	0.218	0.040
3	0.220	0.038
4	0.226	0.032
5	0.230	0.027
6	0.219	0.039
7	0.243	0.015
8	0.220	0.038
9	0.243	0.015
10	0.217	0.041
11	0.244	0.014
12	0.218	0.040
13	0.255	0.002
14	0.234	0.024
15	0.217	0.041
16	0.220	0.037
17	0.242	0.016
18	0.232	0.026
19	0.250	0.008
20	0.244	0.014
21	0.245	0.013
22	0.223	0.035
23	0.229	0.029
24	0.250	0.008
25	0.242	0.016

Compared to the $\overline{X}$ and R charts from Example 1, this control chart is a bit more complicated. The UCL and LCL in the previous variable control chart were constant values and were represented by straight horizontal lines. Owing to the acutely varying sample sizes in this p chart, the UCL and LCL fluctuate. However, it is perfectly acceptable to make evaluations with such a chart that has fluctuating upper and lower control limits. One can still check whether the process is in control or out of control.

Example 2b: Pen Power II

One month after the last analysis, Pen Power again wanted to analyze the number of defective pens produced. However, for this analysis the company made an improvement to simplify chart interpretation: It was more consistent in the amount of pens inspected each day. The data collected are shown in Table 12-4.

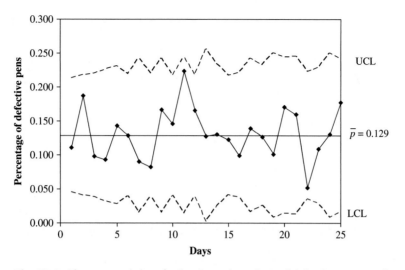

Fig. 12-4. The *p* control chart for Pen Power's analysis of defective pens produced on a daily basis

This problem is very similar to Example 2a. Based on the data given, the total number of rejected pens and total number of pens inspected were calculated, and the process values for the control chart (proportion of defective pens, *p*) were also determined. The results are presented in Table 12-4.

Notice that even though the daily number of inspections still varies, the total numbers of defective pens and pens inspected are the same as in Example 2a—322 defective pens out of 2,500 pens inspected. Thus, $\bar{p}$ is calculated to have the same value as in Example 2a.

However, the upper and lower control limits will not be determined in the same manner as in the previous example. The average sample size will be the same: 2,500 pens in 25 samples, for an average sample size of 100. Whereas the sample sizes in Example 2a varied significantly, from 63 to 144, the data in this second analysis vary much less, ranging from 95 to 105. If the standard deviation were calculated, it would be about 2.87, which is drastically less than the 24.7 in Example 2a. In the first analysis, individual UCL and LCL values were calculated for samples because many sample sizes were not between 80 and 120 pens. In this analysis, however, the samples taken were of more similar sizes, all of which fall within ±20% of the average. As a result, one UCL and one LCL can be calculated for the entire *p* chart. This is done by using the average sample size, $\bar{n}$, in place of *n*, as shown in the following calculations:

$$\text{UCL}_p = 0.129 + \frac{3\sqrt{0.129(1 - 0.129)}}{\sqrt{100}} = 0.229$$

Table 12-4. Daily Counts and Proportions of Defective Pens for Pen Power's Second Analysis

Day (1)	Number of defective pens (2)	Number of pens inspected (3)	p (4)
1	16	97	0.165
2	24	99	0.242
3	12	100	0.120
4	10	104	0.096
5	14	102	0.137
6	16	98	0.163
7	7	98	0.071
8	10	104	0.096
9	13	99	0.131
10	19	101	0.188
11	17	103	0.165
12	21	100	0.210
13	8	103	0.078
14	12	97	0.124
15	16	102	0.157
16	12	101	0.119
17	11	104	0.106
18	12	99	0.121
19	7	96	0.073
20	13	96	0.135
21	12	95	0.126
22	6	105	0.057
23	11	101	0.109
24	9	97	0.093
25	14	99	0.141
Total	322	2500	

$$\text{LCL}_p = 0.129 - \frac{3\sqrt{0.129(1-0.129)}}{\sqrt{100}} = 0.0283$$

At this point, the control chart can be plotted as shown in Fig. 12-5. Because one point is now outside the control limits, the process could be judged as out of control.

Notice that although the total number of rejected pens and total number of inspected pens were the same in both Examples 2a and 2b, their p charts look considerably different. The $\bar{p}$ line remains the same, but the control limits and process values have changed. In Example 2a, the UCL and LCL showed noticeable fluctuations because of the drastically varying sample sizes, but in Example 2b they appear as straight lines because of the relatively similar sample sizes and the fixed sample size used in the calculations. Even the process values do not follow the same pattern because of the difference in sample sizes. When the sample sizes changed,

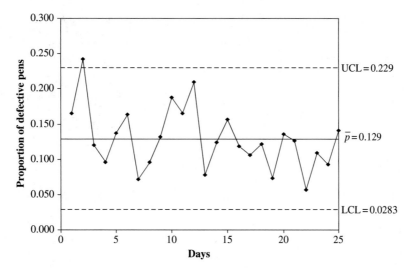

Fig. 12-5. The *p* control chart for Pen Power's second analysis of defective pens produced on a daily basis

the proportions of defective pens changed as well. The effect is visible when comparing the *p* control charts in Figs. 12-4 and 12-5.

12.7.2 The np Chart

The *np* control chart is similar to the *p* chart in that it qualitatively measures defective products. It differs in that, whereas the *p* control chart is concerned with the proportion of defective products, the *np* control chart considers the actual number of defective products. Thus the *np* control chart can be meaningful for data in which the sample size is constant. The *y*-axis of the *np* control chart is the number of defective products.

The *np* control limits can be calculated from Eqs. (12-11) and (12-12). These equations are valid for a spread of 3σ from the centerline. The definition of $\bar{p}$ remains the same as for the *p* chart.

$$\text{UCL}_{np} = n\bar{p} + 3\sqrt{n\bar{p}(1-\bar{p})} \tag{12-11}$$

$$\text{LCL}_{np} = n\bar{p} - 3\sqrt{n\bar{p}(1-\bar{p})} \tag{12-12}$$

Example 3: Pen Power III

Pen Power decides to do a third analysis of its pen production, this time with a constant sample size of 100. Use the data given in Table 12-5 to plot the *np* chart.

Table 12-5. Daily Counts of Defective Pens for Pen Power's Third Analysis

Day	Number of defective pens	Number of pens inspected
1	16	100
2	24	100
3	12	100
4	10	100
5	14	100
6	16	100
7	7	100
8	10	100
9	13	100
10	19	100
11	17	100
12	21	100
13	8	100
14	12	100
15	16	100
16	12	100
17	11	100
18	12	100
19	7	100
20	13	100
21	12	100
22	6	100
23	11	100
24	9	100
25	14	100
Total	322	2500

Solution

Every sample size was equal to 100. Calculate the value of $n\bar{p}$ as $n\bar{p} = 100 \times 0.129 = 12.9$.

Using Eqs. (12-11) and (12-12), compute the upper and lower control limits were computed:

$$\text{UCL}_{np} = 12.9 + 3\sqrt{12.9(1 - 0.129)} = 22.9$$

$$\text{LCL}_{np} = 12.9 - 3\sqrt{12.9(1 - 0.129)} = 2.83$$

The values for $n\bar{p}$, UCL_{np}, and LCL_{np} are plotted as lines on the np control chart shown in Fig. 12-6. This chart also shows the process values—that is, the number of

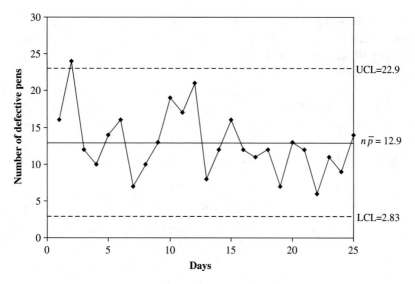

Fig. 12-6. The *np* control chart for Pen Power's third analysis of defective pens produced on a daily basis

defective pens for each day. Again, it is possible to judge the process as out of control, because one point falls outside the control limits.

When Figs. 12-5 and 12-6 (the *p* and *np* charts for the second and third analyses) are compared, we see that the process values show the same trend; the only difference is on the *y*-axis. The *np* chart, which shows the actual number of defective pens, has ordinate values that are *n* times (in this case, 100 times) greater than that of the *p* chart, which shows the proportion of defective pens.

12.7.3 The c Chart

The *c* control chart measures the number of nonconformities, using a constant sample size. It is thus similar to the *np* chart. The *np* chart, however, measures defective products —those that clearly do or do not work; therefore, each unit can only be counted once for the presence or absence of a defect. By contrast, the *c* chart measures nonconformities in products. A single unit can have a number of nonconformities, all of which will be counted for the *c* chart. However, even if a unit has numerous nonconformities, it may not necessarily be defective if it still "works," or functions.

This difference between the *np* chart and the *c* chart (also the *p* chart and the *u* chart) is the reason that different types of distributions are used, which in turn explains why the control limits are calculated in different manners. That is because engineers may be interested in tracking various trends. Whereas the *np* chart uses the binomial distribution, the *c* chart uses the Poisson distribution for rare events.

The term *rare event* is typically used for large samples in which the probability of something occurring is small. In other words, the Poisson distribution is generally used as an approximation of the binomial distribution when the number of samples, n, is large and the probability, p, is small. As a general rule, the Poisson distribution (the c chart and the u chart) is used when the number of samples is at least 50 ($n \geq 50$) and there are no more than five occurrences ($np \leq 5$) (Spiegel and Stephens 2008).

For the c chart, this means that the number of nonconformities should be less than or equal to 10% of the possibilities for nonconformities. Thus, the Poisson distribution is appropriate for the c chart when the number of possibilities for nonconformities to exist is large, but the number of actual nonconformities is small. Eqs. (12-13) and (12-14) for the c chart are based on the Poisson distribution.

$$\bar{c} = \frac{\text{Total nonconformities}}{\text{Number of subgroups}}$$

$$\text{UCL}_c = \bar{c} + 3\sqrt{\bar{c}} \tag{12-13}$$

$$\text{LCL}_c = \bar{c} - 3\sqrt{\bar{c}} \tag{12-14}$$

The application of these equations is relatively straightforward, now that we already know how to make the control charts and plot the data. An example of a problem to which the c chart would be applicable would be determining the number of flaws in a concrete beam or a manufactured door.

12.7.4 The u Chart

The u control chart is a measure of the number of nonconformities with a *varying* sample size. For instance, the u chart would be applicable if a construction company wanted to measure the number of accidents on site each day. The u chart is to the c chart as the p chart is to the np chart: The u and c charts measure the same type of data, but the u chart does not require the constant sample size that the c chart does. The upper and lower control limits are determined by Eqs. (12-15) and (12-16).

$$\bar{u} = \frac{\text{Total nonconformities}}{\text{Total units inspected}}$$

$$\text{UCL}_u = \bar{u} + \frac{3\sqrt{\bar{u}}}{\sqrt{n}} \tag{12-15}$$

$$\text{LCL}_u = \bar{u} - \frac{3\sqrt{\bar{u}}}{\sqrt{n}} \tag{12-16}$$

Just like the p chart, the u chart uses the average sample size, $\bar{n}$, when the sample sizes vary by more than ±20% from the average sample size.

12.8 Control Chart Analysis

Once the correct chart for the given data has been graphed, it must be assessed to acquire information on whether the process is doing well or not—that is, whether it is meeting desired specifications and performing predictably. When using control charts, we term a process either in or out of a state of statistical control, depending on whether or not it is performing satisfactorily. As explained previously, a system or process that is in control is one that has identified and eliminated special (assignable) causes until only common (chance) causes exist.

To assist in the interpretation of control charts, Western Electric Company (WECO) established a set of rules based on probability (Bass 2007). These rules involve visually evaluating the control chart by looking at the process values in comparison with the process average (or centerline) and the upper and lower control limits. When looking at control charts, one is really looking for outlying data points (i.e., variation owing to special causes), unrealistic patterns, and other hints that the process is out of control. In the absence of such findings, one must assume that all variation in the process result from common causes and thus that the process is in control.

The WECO rules are supplemented by the commonly followed practice of evaluating the control chart after partitioning it into six zones, three between the process average and the UCL and three between the process average and LCL. Logically, this would correspond to boundary lines one standard deviation, two standard deviations, and three standard deviations (i.e., the UCL and LCL) above and below the average or centerline. The sections between one standard deviation above and one standard deviation below the centerline are collectively referred to as Zone C. The sections from one to two standard deviations above and from one to two standard deviations below the centerline are collectively referred to as Zone B. Finally, the sections that bookend these zones—between the UCL and two standard deviations above the average, and between the LCL and two standard deviations below the average—are collectively referred to as Zone A. The zones are summarized in Fig. 12-7.

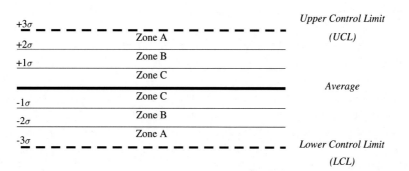

Fig. 12-7. Zones of the control chart

According to Shewhart (1939), the one surefire way to tell that a system or process is out of control is if process values go beyond 3σ (i.e., outside the control limits). This is also stated in the WECO rules ("What are ...", 2017). If even one point falls above the UCL or below the UCL, one may consider the process out of control.

In the normal distribution, 99.73% of all points are within three standard deviations of the average or centerline; there is a 0.27% chance that points will be found beyond 3σ. It is thus considered improbable that points outside the control limits result from common causes. Outlying points are more likely assignable to special causes, which should be eliminated. Therefore, the presence of any point beyond either of the control limits is enough to categorize the process as out of control.

For illustration, refer to the p chart for Example 2b [Fig. (12-5)] and the np chart for Example 3 [Fig. (12-6)]. Each of these control charts has a point above the UCL. Both processes would thus be characterized as out of control and would be investigated to eliminate the cause. Pen Power would have to search for, discover, and resolve the special cause that drove its system out of control. Fig. 12-8(a) shows other examples of a control chart for a process that is out of control.

The remaining WECO rules ("What are ...", 2017) say that a process is out of control when any of the following apply:

- At least two of three consecutive points are located in or beyond Zone A (i.e., beyond 2σ) on the same side of the average. Fig. 12-8(b) shows a control chart with two out of three consecutive points in Zone A below the average.
- At least four of five consecutive points are in or beyond Zone B (i.e., beyond 1σ) on the same side of the process average. Fig. 12-8(c) shows a control chart with four of five consecutive points in and beyond Zone B above the average between days 5 and 12.
- Eight consecutive points are on the same side of the average (other references use nine points instead). See Fig. 12-8(d) for a control chart showing a process that is out of control between days 5 and 12 because there are exactly eight points, one after the other, on the same side of the average.

Brassard (1988) suggested additional suspicious trends that may indicate out-of-control processes:

- Six consecutive points increase or decrease [see Fig. 12-8(e)]. If a process is running in control, variations should be random.
- Fourteen points in a row fluctuate up and down repeatedly [see Fig. 12-8(f)]. In a process that has random fluctuations but is in control, the fluctuations should not display a discernible pattern.

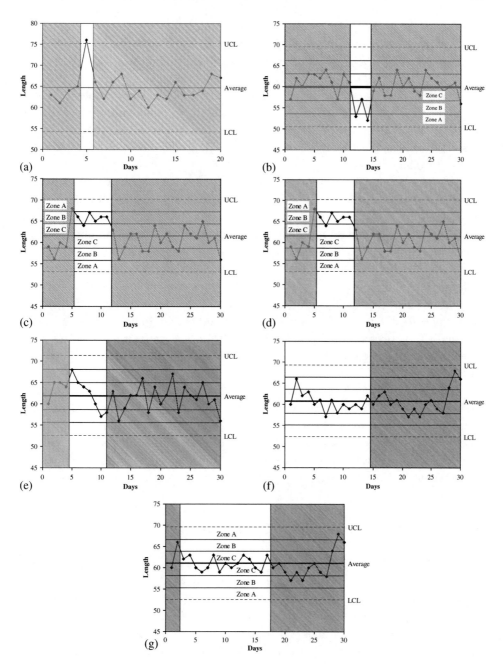

Fig. 12-8. (a) One point above the UCL; (b) Two of three consecutive points in Zone A below the average; (c) Four of five consecutive points in and beyond Zone B above the average; (d) Eight consecutive points above the average; (e) Six consecutive decreasing points; (f) Fourteen consecutive points going up and down; (g) Fifteen consecutive points within Zone C

- Fifteen successive points are in Zone C, either above or below the centerline [see Fig. 12-8(g)]. In a normally distributed dataset, it would be quite unusual to see 15 consecutive points with very low standard deviations.

Brassard (1988) states that if a control chart shows a considerable display of systematic trends, there is a systemic problem causing an out-of-control process that, therefore, needs attention. The process may go out of control for any number of reasons—workers, machines, materials, environment, communication, calibration, design, or many others.

Shewhart's rule ("Shewhart Control Chart Rules," 2017; Shewhart 1939) that data points falling outside 3σ signify that a process is out of control is considered the most effective. The other rules and suggestions may give signs that a process is out of control, but they do not dictate as much. In fact, using a number of these trends to interpret control charts may even lead to false alarms (Drain 1997). One could stop production in such cases only to discover that there is nothing wrong with the system. Simply put, suspicious trends give reason to look into the cause of the pattern before classifying the control chart. If special causes were indeed responsible for the trend, then such causes should be eliminated before repeating the analysis to check that the process is in control. The charts in Fig. 12-8 a-g show processes that may be out of control and should be further investigated for special causes.

If the analysis is performed correctly and the process is deemed out of control, the next step is to find and eliminate the special causes of variation. All possible sources of variation, such as people, equipment, or environment, must be investigated until the special causes are discovered. All in all, control charts help eliminate or reduce the uncertainties in the system, permitting managers an analytical tool that can mitigate and control risks.

12.9 Summary, Analysis, and Conclusion

Control charts are useful for tracking performance and, thus, risk. If the process is inconsistent, or out of control, control charts can help identify causes of variation. Ideally, after a special cause is known, it can and will be removed to improve consistency. Once the process is in control, the data collected will be more dependable and can be used as a barometer of future performance.

It is important to note that control limits are not the same as specification limits. Control limits are objective; they are statistically calculated and represent the actual operating conditions of the process. Specification limits, by contrast, are subjective and indicate the desired conditions of the process. If the actual operating conditions of a process are consistent, the process may be considered in control, but this is not an indication that it meets specifications;

it means only that the machines or people are performing as predicted. If the process consistently fails to meet requirements, it may be in control, but it does not meet specifications. To truly implement TQM and improve the quality of a process or system, one must balance control against the ability to meet the requirements of customers (Brassard 1988). Process capability, which is not covered in this text, typically uses control charts to assess how standards are or are not being met.

Control charts are not the be-all and end-all of TQM. They simply provide somewhere to start. Control charts show the actual operating conditions of the process, particularly emphasizing how process values are distributed within three standard deviations of the process average. They may lead to the discovery and resolution of special causes of variation; they may also prevent unnecessary modifications. Although these charts will certainly provide steps toward improving quality, there is more to do. Remember that the objective of TQM is not only to improve, but to *continuously* improve.

The topic of control charts can be simplified to the following method:

1. Determine the process or system to be evaluated, and identify the type of data to express it in. Based on the type of data, choose the appropriate type of control chart to use.
2. Calculate the process average, UCL, and LCL; then graph these three lines along with the process values from the data.
3. Evaluate the process or system (i.e., classify it as in control or out of control). If it is out of control, find and remove special causes until only common causes remain and the process or system is in control.
4. Once the process or system is in control, continue to pursue TQM, and continue to improve the quality of the process or system.

By using this method, operation managers can get a handle on risk and exercise better control over the process.

12.10 Exercises

Create control charts for each of the following situations. Determine the state of control.

a) A new company has begun the process of making clock radios. They would like to employ the principles of TQM and start making improvements to their process. They took samples of their product and counted the number of clock radios that did not work.

Day	Number of defectives	Number inspected
1	24	75
2	22	75
3	15	75
4	13	75
5	5	75
6	21	75
7	8	75
8	16	75
9	23	75
10	10	75
11	11	75
12	24	75
13	6	75
14	11	75
15	5	75

b) A construction company would like to know how many injuries and near-misses occur on the job. The potential for accidents depends on what sort of activity occurs at a given time. The company would like to make a control chart to reduce its risk in the future.

Month	Number of injuries/near-misses	Number of potential injury opportunities
1	33	1,200
2	37	985
3	38	1,070
4	31	1,000
5	21	1,220
6	40	1,250
7	36	1,190
8	32	930
9	41	1,070
10	20	960
11	28	1,100
12	31	1,275

c) A researcher collected the following data on the water velocity in a pipe over ten hours.

	Flow velocity (ft/s)				
Hours	1	2	3	4	5
1	2.4	2.7	2.1	2.5	2.7
2	2.4	2.6	2.5	2.3	2.3
3	2.8	2.4	2.6	2.5	2.2
4	2.1	2.7	3.0	2.2	2.9
5	2.2	2.0	2.5	2.8	2.1
6	3.0	2.1	3.2	2.7	2.2
7	2.4	2.6	2.9	2.6	2.5
8	2.6	2.3	2.3	2.9	2.2
9	2.5	2.8	2.7	2.7	3.1
10	2.9	3.0	2.6	2.2	2.9

d) The following data were collected for a toy robot manufacturer to see whether its process is consistent.

Day	Number of rejects	Number inspected
1	19	87
2	20	69
3	18	63
4	17	73
5	11	78
6	9	83
7	13	72
8	6	81
9	14	78
10	17	82
11	12	71
12	11	78
13	15	83
14	8	72
15	13	81
16	18	78
17	12	82
18	15	72
19	12	66
20	18	73

e) A student who is doing a survey must fill out paperwork that requires typing in 1,000 message fields every week. The number of errors (misspellings, typos, wrong information, etc.) made were counted as shown.

Week	Number of errors	Number of opportunities
1	17	1,000
2	31	1,000
3	28	1,000
4	41	1,000
5	50	1,000
6	36	1,000
7	24	1,000
8	29	1,000
9	18	1,000
10	52	1,000
11	41	1,000
12	19	1,000
13	25	1,000
14	32	1,000
15	51	1,000
16	47	1,000

References

Akpolat, H. (2004). *Six sigma in transactional and service environments*, Gower, Aldershot, U.K.

Analyse-it Software, Ltd. "Shewhart control chart rules." <https://analyse-it.com/docs/user-guide/processcontrol/shewhartcontrolchartrules> (Mar. 27, 2017).

Bass, I. (2007). *Six sigma statistics with excel and minitab*, McGraw-Hill, New York.

Best, M., and Neuhauser, D. (2006). "Walter A. Shewhart, 1924, and the Hawthorne factory." *Qual. Saf. Health Care*, 15(2), 142–143.

Brassard, M. (1988). *The memory jogger: A pocket guide of tools for continuous improvement*, Goal/QPC, Salem, NH.

DataNet. (2017). "What are the western electric rules." Western Electric – A brief history, *Western Elec. News*, 11(2), Western Electric, Rossville, GA, 1913. <http://www.winspc.com/14-datanet-quality-systems/support-a-resources/179-what-are-the-western-electric-rules> (May 10, 2017).

Deming, W. E. (2000). *The new economics: For industry, government, education*, 2nd Ed., MIT Press, Cambridge, MA.

Drain, D. (1997). *Statistical methods for industrial process control*, Chapman and Hall/CRC, Boca Raton, FL.

Ehrlich, B. H. (2002). *Transactional six sigma and lean servicing*, CRC Press, St. Lucie, Boca Raton, FL.

"Shewhart Control Chart Rules." Analyse-it Software, Ltd., Leeds, UK <https://analyse-it. com/docs/user-guide/processcontrol/shewhartcontrolchartrules> (Mar. 27, 2017).

Shewhart, W. (1939). *Statistical method from the viewpoint of quality control*, Dover Publications, Mineola, NY.

Skymark. (2017). "Walter Shewhart—The grandfather of total quality management." <http://www.skymark.com/resources/leaders/shewart.asp> (Mar. 30, 2017).

Smith, J. (2011). "Management: The lasting legacy of the modern quality giants, quality magazine." <http://www.qualitymag.com/articles/88493-management–the-lasting-legacy-of-the-modern-quality-giants> (Mar. 28, 2017).

Spiegel, M. R., and Stephens, L. J. (2008). *Schaum's outline of theory and problems of statistics*, 4th Ed., McGraw-Hill, New York.

Walesh, S. G. (2000). *Engineering your future: The non-technical side of professional practice in engineering and other technical fields*, 2nd Ed., ASCE, Reston, VA.

Western Electric. (2017). "Western electric history." <http://www.beatriceco.com/bti/porticus/bell/westernelectric_history.html#Western%20Electric%20-%20A%20Brief%20History> (Mar. 27, 2017).

Xie, M., Goh, T. N., and Kuralmani, V. (2002). *Statistical models and control charts for high-quality processes*, Kluwer Academic, Norwell, MA.

CHAPTER 13

Economic Order Quantity for Inventory Control

13.1 Introduction

The success of a company or business relies on many factors. One is its ability to efficiently manage product inventory, especially when inventory is in the millions of dollars and occupies space over many warehouses in many cities. Efficient management ensures that the costs of acquiring and storing inventory are minimized while the number of products in inventory is maximized. While enormous costs will accrue if inventory is ordered each time it is needed rather than stored, it is also uneconomical to store more products than there is space available. Storing too much inventory results in lost dollars, while storing too little may result in lost opportunities. These risks can be better managed by using proven inventory control methods. One such method is the *economic order quantity*.

Each cost is based on the inherent risk of doing business. When it comes to inventory management, risks are ever present because of the inability to make exact predictions of demand, as well as cost fluctuations stemming from outside factors such as fuel and labor. It is almost impossible to eliminate all risks involved with inventory management, but they can be significantly reduced within the strict scope of determining an optimal order quantity. This chapter will illustrate an inventory management technique to contain risks, as well as how to calculate the typical costs involved.

13.2 Costs Associated with Carrying Inventory

13.2.1 Ordering Costs

The first type of cost typically encountered in inventory procurement is the cost of placing an order. In the real world of business, the process of ordering products is not free.

To place orders and receive products, a business must

1. go through the process of selection, which can consume many hours or even days;
2. review justifications for purchase from engineers or officers;
3. have accountants prepare purchase orders that then go through an approval process;
4. place orders by making phone calls, sending faxes or emails, or using a courier service; and then
5. arrange to receive deliveries.

All of these steps consume worker hours and incur business expenses such as overhead costs. Higher or lower costs may be incurred depending on businesses' individual situations and the items or services procured.

Ordering costs are normally considered fixed costs, because the quantity of paperwork, for instance, is assumed to be the same for a large order as for a small one. In any event, an average of all costs experienced will be used as the fixed ordering costs. Therefore, we can determine total annual ordering costs by simply multiplying the *fixed cost* of placing and receiving an order (F) by the *number of orders* placed per year (N). We then have Eq. (13-1) for *total ordering costs* (TOC):

$$TOC = F \times N \tag{13-1}$$

For example, assume that a company incurs a fixed ordering cost of \$150 ($F = \$150$) and places monthly orders ($N = 12$ orders per year). Eq. (13-2) then gives us the TOC:

$$TOC = F \times N = \$150 \times 12 = \$1,800 \tag{13-2}$$

It is of interest to note that in certain cases it is useful to add another term, *shipping and receiving costs*, to the ordering-cost component of the inventory cost model. It is beneficial to include this term when there are economies of scale in shipping—that is, when it costs less per unit to ship an item if the order size is large. However, this often is not the case in industrial merchandise, so that shipping cost is not directly related to order size and is simply the shipping cost per item multiplied by the number of units ordered. In this scenario, a separate term for shipping and receiving costs may be ignored. For large industrial items such as turbines or aircraft wings, though, large wagons may be needed for transport and have a large cost of their own. Note that it is customary to see shipping and receiving costs added to the cost of the item and not included as part of the ordering costs.

13.2.2 Carrying Costs

The next typical cost encountered is the cost of carrying inventory, which is generally directly proportional to the number of products held in inventory. In most cases, the inventory carried by a company is determined by the frequency at which inventory is ordered. The higher the ordering frequency, the less inventory is required. If we define the product units sold by the company as S and use N once again for the number of orders placed per year, we arrive at Eq. (13-3) for the average inventory, A, carried by the company:

$$A = \frac{S/N}{2} \qquad (13\text{-}3)$$

Hence, fewer orders per year will result in larger inventories, and vice versa. For example, assume that a company sells 60,000 units per year and places orders every month. The average inventory will then be found by Eq. (13-4):

$$A = \frac{S/N}{2} = \frac{60,000/12}{2} = 2,500 \text{ units} \qquad (13\text{-}4)$$

Using the above average inventory, along with the fact that the company purchases inventory at a price, P, of \$10 per unit, the average inventory value can be calculated by Eq. (13-5):

$$(P) \times (A) = (\$10) \times (2,500) = 25,000 \text{ units} \qquad (13\text{-}5)$$

The carrying costs involved can be further broken down into subcategories (capital, storage costs, insurance, utility expenses, depreciation and obsolescence, etc.). Storage costs can be further broken down to include costs related to space available, security, taxes, and so forth; utility costs involve any air conditioning, electricity, and water[1] necessary for storage. These costs can be summed up to arrive at a factor known as the *total carrying cost* (TCC).

For example, assume that a company has a capital cost of 10% of the average inventory value, or \$2,500. Then let's say that storage costs along with utility costs equal \$1,500; insurance costs \$500; and depreciation and obsolescence cost \$1,000, for a total of \$2,500 + \$1,500 + \$500 + \$1,000 = \$5,500. Therefore, the percentage cost of carrying the 25,000 units inventoried at any one time is \$5,500/\$25,000 = 0.22 = 22%. We can define this *percentage carrying cost* as *C*. We can then find the total carrying cost by taking the percentage cost, multiplied by the price per unit (P), multiplied by the average inventory (A) [Eq. (13-6)]:

$$\text{Total Carrying Cost (TCC)} = C \times P \times A \qquad (13\text{-}6)$$

Applying this equation to our example, we obtain Eq. (13-7):

$$\text{TCC} = 0.22 \times \$10 \times 2,500 = \$5,500 \qquad (13\text{-}7)$$

---

[1]Electronic items may require air conditioning; landscaping products may require water; . . . and so on.

It is of interest to note that average values of percentage carrying cost, C, will normally be between 15% and 25% in large industrial organizations run professionally.

13.2.3 Total Inventory Cost

We have so far determined two of the major costs associated with inventory management, ordering and carrying costs. No other costs are assumed to exist, because all are included in these two types of costs. By summing the two, we can develop Eq. (13-8) for the *total inventory cost* (TIC):

$$\text{Total Inventory Cost (TIC)} = \text{TOC} + \text{TCC} = (F \times N) + (C \times P \times A) \quad (13\text{-}8)$$

If we rearrange the average inventory relationship derived in Eq. (13-3) and solve for N, we find that $N = \frac{S}{2A}$. Substituting this into Eq. (13-8), we have Eq. (13-9):

$$\text{TIC} = F \times \left(\frac{S}{2A}\right) + (C \times P \times A) \quad (13\text{-}9)$$

We can also recognize that the size of each order, Q, is merely twice the average inventory: $Q = 2A$. Looked at differently, the inventory goes from Q to 0, and so the average inventory is $A = (Q + 0)/2 = Q/2$. Substituting this into Eq. (13-9) gives us Eq. (13-10):

$$\text{TIC} = \frac{(F \times S)}{Q} + \left(C \times P \times \frac{Q}{2}\right) \quad (13\text{-}10)$$

Even though we have developed a relationship to determine the total inventory costs a company will experience, that equation alone cannot show us how to properly manage our inventory. If too much or too little is ordered, the company will suffer from increased costs. To ease this uncertainty, we need to find an *economic order quantity*.

13.3 Economic Order Quantity

We notice that certain costs increase with larger order sizes, whereas other costs increase with smaller order sizes. To understand this, let's look first at the total ordering costs. If we order very frequently—let's say daily—we will most likely be ordering in small quantities, which results in a large associated ordering cost over the year because of the large, repetitive costs of simple ordering.

The total carrying costs behave in a linear fashion. If we have small order sizes, we will have only a small inventory stock, resulting in a small associated cost.

However, if we purchase in very large quantities, we will in turn have larger inventories that increase the carrying cost. Since carrying cost is directly related to the amount of inventory on hand, having a larger order size would mean paying more for such expenses as storage space, depreciation, and insurance. Obsolescence would also be higher because items would sit in inventory longer.

Summing the TOC and TCC curves gives us the TIC curve. By examining this curve in Fig. 13-1, we notice that the TIC initially decreases as order size increases, but after a certain point it begins to increase again. The point on the curve where the total inventory cost is minimal is known as the *economic order quantity* (EOQ) (Waters 2003; Brigham and Gapenski 1985).

13.3.1 Derivation of EOQ

To determine the EOQ, we differentiate the TIC [Eq. (13-10)] with respect to the order size, Q and set the resulting derivative equal to zero [Eq. (13-11)]:

$$\frac{\partial(\text{TIC})}{\partial Q} = -\frac{F \times S}{Q^2} + \frac{C \times P}{2} = 0 \qquad (13\text{-}11)$$

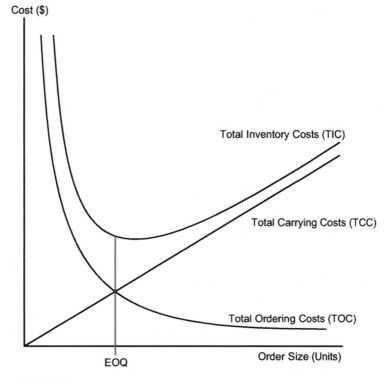

Fig. 13-1. Inventory cost curve

Now, rearrange the equation and solve for Q to get Eq. (13-12):

$$\frac{C \times P}{2} = \frac{F \times S}{Q^2} \qquad (13\text{-}12)$$

From which comes Eq. (13-13):

$$Q^2 = \frac{2F \times S}{C \times P} \qquad (13\text{-}13)$$

and then Eq. (13-14):

$$Q = \text{EOQ} = \sqrt{\frac{2F \times S}{C \times P}} \qquad (13\text{-}14)$$

where Q = economic order quantity, or optimal quantity to be ordered each time an order is placed; F = fixed cost associated with placing and receiving a single order; S = number of units sold annually; C = cost of carrying inventory, expressed as a percentage; and P = price per unit purchased by the company.

13.3.2 Assumptions of the EOQ Model

The EOQ model is derived on the basis of a few assumptions.

- The first is that there is repetitive ordering. Once the company notices that inventory has been used up, another order will be immediately placed to replenish the stock. This will happen every time the inventory level depletes.
- The second assumption is that there is a constant demand for the product which occurs at a known rate. For example, if the company foresees a demand of 60,000 units per year, the demand during m months will be $\frac{60,000\,m}{12}$ units.
- The third assumption is that there is a constant lead time, defined as the time between when an order is placed and when it arrives—if an item has a lead time of two weeks, it will take two weeks for the order to arrive once it is placed. It also assumes there are no unforeseen delays in the delivery of the order, such as transportation or production delays (Winston 1991; Waters 2003). These simplifications make the calculations easier without resorting to complex calculus, and the results from more sophisticated stochastic analysis are often no better than those from this simple method.
- The final assumption of the model is that continuous ordering is employed. Inventory is constantly reviewed, and orders can be placed at any given time.

We will illustrate the usage of the EOQ model with the following example.

13.4 EOQ Example

SpeedRun is a shoe distribution company. Its top product is a high-performance running shoe. Annually, the company expects to sell 13,000 pairs. The percentage carrying cost of holding these shoes in inventory is 20% of the inventory value. SpeedRun purchases the running shoes from the manufacturer at a purchase price of $28.89 per pair. The fixed cost of placing an order is taken to be $500.

Substituting the given data into equation (13-14), an EOQ of 1,500 pairs is obtained quite easily [Eq. (13-15)]:

$$\text{EOQ} = \sqrt{\frac{2 \times F \times S}{C \times P}} = \sqrt{\frac{(2)(\$500)(13,000)}{(0.20)(\$28.89)}} = \sqrt{2,249,913.5} = 1,500\,\text{pairs}$$

$$(13\text{-}15)$$

Now that we know the EOQ, we can also find how often SpeedRun must place orders. This is done simply by dividing the EOQ by the annual demand, S, from Eq. (13-3), giving us Eq. (13-16):

$$N = \left(\frac{1,500\,\text{pairs}}{13,000\,\frac{\text{pairs}}{\text{year}}}\right)\left(52\,\frac{\text{weeks}}{\text{year}}\right) = 6\,\text{weeks} \qquad (13\text{-}16)$$

Which amounts to placing $S/Q = 52/6 = 8.67$ orders per year. SpeedRun will thus place orders every 6 weeks at a quantity of 1,500 pairs in order to maintain an adequate inventory while minimizing total costs. This will roll over into the next year.

If we wish to determine the total inventory cost incurred by ordering at EOQ, we simply plug in our EOQ value into the total inventory cost equation to find the total cost [Eq. (13-17)]:

$$\text{TIC} = \frac{(F \times S)}{Q} + C \times P \times \left(\frac{Q}{2}\right) \qquad (13\text{-}17)$$

Therefore, we find SpeedRun's total inventory cost with Eq. (13-18).

$$\text{TIC} = \frac{(\$500 \times 13,000)}{1,500} + 0.20 \times \$28.89 \times \left(\frac{1,500}{2}\right) = \$8,666.83 \qquad (13\text{-}18)$$

13.5 Reorder Point

Practically speaking, orders do not arrive the instant they are placed. Therefore, we cannot wait until our inventory level is entirely depleted before placing another order. Management must neither take unnecessary risks nor forget the reordering schedule. We must consider the lead time, or the time it takes for an order to arrive once it is placed.

There are two cases we must consider. The first case is where the lead time in years, L, multiplied by demand per year, S, is less than the EOQ (i.e., $L \times S <$ EOQ). The second case is where the lead time multiplied by demand is greater than the EOQ (i.e., $L \times S >$ EOQ).

13.5.1 Case where L × S < EOQ

Consider the following for the first case. If SpeedRun sells 13,000 pairs of shoes per year and there is a two-week lead-time for the order to arrive, we get Eq. (13-19):

$$L \times S = \left(\frac{2\,\text{weeks}}{52\,\frac{\text{weeks}}{\text{year}}}\right)\left(13{,}000\,\frac{\text{pairs}}{\text{year}}\right) = 500\,\text{pairs} < EOQ\,(= 1{,}500\,\text{pairs}) \qquad (13\text{-}19)$$

where L is in years and S is in pairs per year.

Alternatively, because 13,000 shoes per year equals 250 shoes per week, SpeedRun must place an order when there is an inventory level of $250 \times 2 = 500$ pairs of shoes. If the order is placed at this precise inventory level, the inventory level will reach 0 pairs just as the new shoe order arrives. If the order is placed too late or too early, there will be a shortage or surplus in inventory, respectively. Fig. 13-2 shows

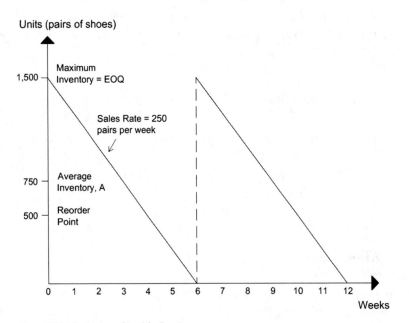

Fig. 13-2. Inventory level behavior

the order amount, Q; the average inventory, A; the order period, 6 weeks; the sales rate, 250 shoes per week; and the reorder point, 2 weeks.

13.5.2 Case where L × S > EOQ

For the second case, where $L \times S > $ EOQ, let's say that SpeedRun experiences a lead time of 10 weeks from the manufacturer. This would result in Eq. (13-20):

$$L \times S = \left(\frac{10 \, \text{weeks}}{52 \, \frac{\text{weeks}}{\text{year}}} \right) \left(13,000 \, \frac{\text{pairs}}{\text{year}} \right) = 2,500 \, \text{pairs}$$

$$> EOQ \, (= 1,500 \, \text{pairs}) \tag{13-20}$$

This breaks the EOQ mold. To alleviate the issue of reordering with a long lead time, our first order must be placed 10 weeks prior to when it is needed. We also note from Eq. (13-16) that, in the optimal model, orders should be placed every 6 weeks. Thus, our first order will need to be placed at a time of -10 weeks, the second order at -4 weeks, the third at $+2$ weeks, and so on. Our first order will then arrive at week 0, and the inventory at that time will be 1,500 pairs. We can now find the reorder point by examining the inventory level when the next order is placed at $t = -4$ weeks. After $t = 0$ weeks, the next order will be placed at week $+2$. In two weeks, the inventory level will have declined by $\frac{2 \, \text{weeks}}{52 \, \text{weeks/year}} \times 13,000 \frac{pairs}{year} = 500 \, pairs$, resulting in a reorder point of $1,500 - 500 = 1,000$ pairs.

Therefore, whenever our inventory level reaches 1,000 pairs, we must place another order to ensure that our inventory does not become depleted. The reorder point can also be calculated in this case by simply taking the remainder of $L \times S$ divided by EOQ. For instance, $\frac{2,500}{1,500} = 1 \, remainder \, 1,000$, so 1,000 pairs will be the reorder point, just as we determined previously. In other words, a four-week cushion will be necessary.

13.6 Sensitivity of EOQ Model

The total inventory cost is not extremely sensitive to slight deviations from the EOQ. Examining the TIC curve in Fig. 13-1, notice that the slope of the curve near the EOQ point is relatively flat to both the left and right sides. For example, if we were to order 1,400 pairs of shoes instead of the EOQ of 1,500 pairs, our TIC would be \$8,687.46[2], a difference of a mere \$20.63. On the other hand, if we ordered 1,600 pairs of shoes instead of 1,500, our TIC would be \$8,684.90[3], just \$18.07 greater than optimal. In essence, this low sensitivity allows us to have a safety cushion for uncertainties that might arise during a real-life application (Tersine

---

[2]TIC $= FS/Q + CPQ/2 = (500)(13,000)/(1,400) + (0.20)(\$28.89)(1,400)/2 = \$8,687.46$.
[3]TIC $= FS/Q + CPQ/2 = (500)(13,000)/(1,600) + (0.20)(\$28.89)(1,600)/2 = \$8,684.90$.

1994). It could also be said that nature plays a role, albeit a limited one, in preserving the extra expenses owing to small variations and uncertainties.

13.7 Average Inventory

Once an order is received, the inventory will reach a maximum value of $Q(= EOQ)$. Following our assumption that units are depleted at a constant rate, we can recognize the sales rate as 250 pairs per week (13,000 pairs/52 weeks), thus reducing the inventory level by 250 pairs each week. Ultimately, the inventory level will vary from 1,500 pairs (when an order is received) down to 0 pairs (the instant before another order arrives). The average inventory level will be one-half of the EOQ, 750 pairs. Integrating the cost per shoe of $28.89, SpeedRun will have an average inventory investment of $28.89 × 750 = $21,667.50. This value will range from $43,335 when inventory is at a maximum down to $0 when inventory is entirely depleted (Winston 1991).

13.8 Weaknesses of the EOQ Model

The EOQ model is a powerful tool in inventory management, but it comes with its own sets of limitations. We will discuss a few of them in detail.

The first limitation is that the model will often present inconvenient order quantities. The EOQ will not always give a whole number; more often than not, it will suggest fractional quantities for items that are only available as discrete units. This becomes a concern when the items in question possess a high monetary value while occurring in small quantities. For example, EOQ may suggest that 1.6 aircraft should be ordered. It is not possible to split aircraft to achieve such a value, and large consequences will occur if one rounds down to 1 or up to 2. Ordering one aircraft for an airline could mean that inventory would be depleted too soon, resulting in increased ordering costs; ordering two aircraft could mean that there would be an excess of inventory and therefore increased carrying costs, not to mention the loss of revenue because of the lack of full use of the second aircraft. This is a somewhat extreme example, but it serves to drive home the point.

Another limitation is also related to inconvenient order quantities. Suppliers are normally unwilling to split standard-size packages. To illustrate this, let's assume that the EOQ for an order of rice is 16 pounds. Rice is packaged in standard sizes of 5-, 10-, 20-, 50-, and 100-pound bags. If a company were to strictly follow the EOQ, it would have to choose among ordering one 10-pound bag of rice, one 10-pound bag and one 5-pound bag, two 10-pound bags, one 10-pound bag and two 5-pound bags, or one 20-pound bag—and must evaluate the increased costs associated with all alternatives.

The EOQ model may also suggest that orders be placed at inconvenient points. It is convenient for companies to place orders at regular intervals, be it weekly, bimonthly, or monthly, for example. When the EOQ model is used, it may suggest that orders be placed at irregular times, such as every 5.25 weeks. This may be difficult for companies to efficiently schedule amongst their numerous other responsibilities, not to mention deciding what to do with the fractional week (Thomas 1969).

Then there is the issue of transportation capacity. Real-world deliveries are made using vehicles such as trucks, ships, trains, and planes, all of which have fixed carrying capacities. For example, a truck has the capacity to haul 12 tons of material, but our EOQ calls for 15 tons of material. In order to meet the EOQ, two trucks will be required to make the delivery, resulting in double the delivery cost.

The EOQ model's assumption that demand is constant can lead to an obsolescence of stock, besides being fraught with danger. Applying the EOQ without proper regard to falling demand, for instance, can lead to a high inventory level, meaning excessive carrying costs. To alleviate this problem, demand should be evaluated regularly and the EOQ adjusted accordingly. In the end, it is important to address and structure uncertainty to fit the EOQ model.

13.9 Safety Stock

One of the underlying assumptions of the EOQ model is that there is a constant demand for the product. In this section we will relax this constraint in an effort to mold our model more closely to the real world. Practically speaking, a company will not have constant demand and may experience a sudden demand surge or slowdown. During slowdowns, the company is left straddled with inventory without revenue through sales, while during surges, in combination with lead times for delivery, the company may not be able to replenish its inventory in time, and thus could lose on potential profits.

We can illustrate this using the SpeedRun example again. Let's say that instead of the normal weekly demand of 250 pairs, the demand doubles to 500 pairs. If operations continue using the same reorder point of 250 pairs, inventory will be depleted a week before a new order arrives. Having a depleted inventory can lead to either lost sales or back orders, which may or may not be accepted, depending on the nature of the customer. If the customer is not in a rush, he or she may agree to a back order and wait the extra time to receive the order. Alternatively, the customer may not want to wait and will go elsewhere for the same or similar product, which would result in lost sales.

Lost sales can be minimized by using the concept of *safety stock*. Safety stock simply means extra stock kept on hand in case inventory is depleted before the next order arrives. Once a company decides to carry safety stock, the next step is to decide how much. As discussed previously, there are costs associated with holding stock, and that includes safety stock: the more safety stock, the higher the carrying costs. For instance, if SpeedRun decides to carry a safety stock of 750 pairs of shoes, the carrying cost will increase from the original $4,333.50 to $8,667.00. This can be illustrated with Eq. (13-21), with safety stock denoted as SS (Brigham and Gapenski 1985; Toomey 2000):

$$\text{Original Carrying Cost} = C \times P \times \left(\frac{Q}{2}\right) = (0.2)(\$28.89)\left(\frac{1,500}{2}\right) = \$4,333.50$$

$$\text{Safety Stock Carrying Cost} = (C)(P)(SS) = (0.2)(\$28.89)(750) = \$4,333.50$$

$$\text{Total Carrying Cost} = \text{Original Carrying Cost} + \text{Safety Stock Carrying Cost}$$

$$= \$4,333.50 + \$4,333.50 = \$8,667.00 \tag{13-21}$$

In other words, carrying a safety stock that is *half* the EOQ *doubles* the total carrying cost.

13.9.1 Setting the Safety Stock Level

Before a company begins to stockpile safety stock, a proper analysis should be carried out based on the probability distribution of a demand forecast. Consider the sales probability distribution in Table 13-1.

It shows the different demand probabilities that SpeedRun will face over a six-week period, which will affect the amount of time needed to deplete an order of 1,500 pairs of shoes. Expected sales—1,500 pairs—are the same constant demand that we worked with previously.

We determined earlier that SpeedRun's annual carrying cost is 20% of the inventory value. Multiplying this annual cost by the inventory value—the price paid per unit by the company—we determine that the annual carrying cost per unit is $(0.20) \times (\$28.89) = \$5.778$. It can be further established that the carrying cost per unit for a six-week period is $(6/52) \times (\$5.778) = \$0.667$.

Table 13-1. Six-Week Sales Probability Distribution

Probability	Sales (pairs)
0.1	1,000
0.2	1,250
0.4	1,500
0.2	1,750
0.1	2,000
$\sum$ prob. = 1.0	Expected sales = 1,500

The next step is to estimate the cost that will be incurred if a shortage occurs. To simplify this example, it will be assumed that half of SpeedRun's customers will accept back orders, but the other half will not purchase anything during the six-week period. Assuming that SpeedRun sells each pair of shoes for $50, each pair that is short of demand will cause SpeedRun to incur a lost profit equal to $(0.5) \times (\$50 - \$28.89) = \$10.56$. Table 13-2 shows an example of calculating the costs associated with different safety-stock levels, and analyzes the costs of employing different safety-stock levels in association with the six-week sales probability distribution.

The shortage amount is determined by subtracting the sales column from the expected sales of 1,500 pairs. So for expected sales of 1,000 pairs of shoes, there will be no shortage, but for expected sales of 1,750 pairs, there will be a possible shortage of 250 pairs of shoes. Shortage cost is calculated by multiplying the shortage amount by the lost profit value of $10.56. The expected shortage cost is then calculated by multiplying the shortage cost by its probability of occurrence. The safety-stock carrying cost is not a probabilistic number and is calculated simply by multiplying the individual-unit carrying cost for the 6-week period ($0.667) by the amount of safety stock carried. Finally, the expected total cost is the sum of the expected shortage cost and the safety-stock carrying cost.

Using a safety-stock level of 0, we would expect a shortage cost of $1,056.00. If we increase our safety stock to 250 pairs, our shortage cost will decrease to $264.00—but a safety-stock carrying cost of $166.75 will be incurred, leading to an expected cost of $430.75. Further increasing our safety stock to 500 pairs would eliminate the expected shortage cost but incur a higher safety-stock carrying cost of $333.50. If SpeedRun decides to carry a safety stock greater than 500 pairs, there will be even higher expected total costs. In this case, the optimum safety stock to carry would be 500 pairs of shoes, as this will provide the lowest expected total cost to the company. These calculations are presented in Table 13-2.

Note that the safety-stock analysis will be only as good as a company's estimate of the sales probability distribution and lost profits. There is also an assumption that lost profit in one period will not carry over to subsequent periods. In real-life situations, however, there may be a loss of goodwill if the company is not able to provide customers with the desired product at the time of demand, possibly resulting in the customer permanently taking his business elsewhere. Ultimately this could result in a long-term reduction of sales. This is just one example of the complexity of determining inputs into the EOQ model. The model itself is simple to use—the human behavior aspect governing the inputs is the complex part (Waters 2003).

13.10 EOQ with Back-Ordering

Instead of safety stock, an alternative method of dealing with shortages is back-ordering. In certain situations it may be desirable to allow for back-ordering as an economical alternative to shortages. Back-ordering implies that an order was placed

Table 13-2. Safety Stock Analysis

Safety stock [1]	Sales [2]	Probability [3]	Shortage [4]	Shortage cost $10.56 \times [4]$ [5]	Expected shortage cost $[3] \times [5]$ [6]	Safety-stock carrying cost $0.667 \times [1]$ [7]	Expected total cost $[6] + [7]$ [8]
0	1,000	0.1	0	0	0		
	1,250	0.2	0	0	0		
	1,500	0.4	0	0	0		
	1,750	0.2	250	$2,640	$528		
	2,000	0.1	500	$5,280	$528		
				Total	$1,056.00	$0.00	$1,056.00
250	1,000	0.1	0	0	0		
	1,250	0.2	0	0	0		
	1,500	0.4	0	0	0		
	1,750	0.2	0	0	0		
	2,000	0.1	250	2640	264		
				Total	$264.00	$166.75	$430.75
500	1,000	0.1	0	0	0		
	1,250	0.2	0	0	0		
	1,500	0.4	0	0	0		
	1,750	0.2	0	0	0		
	2,000	0.1	0	0	0		
				Total	$0.00	$333.50	$333.50
750	1,000	0.1	0	0	0		
	1,250	0.2	0	0	0		
	1,500	0.4	0	0	0		
	1,750	0.2	0	0	0		
	2,000	0.1	0	0	0		
				Total	$0.00	$500.25	$500.25

by a customer, the item was not in current inventory, the order that was placed by the customer was not withdrawn, and the order is filled once the next shipment arrives (Gupta and Khanna 2011).

However, back-ordering does incur costs, such as lost business, the cost of placing special orders, and loss of future goodwill. These shortage costs can be combined into a single term, s', which represents the shortage cost per unit for a single year.

Eqs. (13-22) through (13-25) will be useful when faced with a situation where back-ordering is allowed; they assume a zero lead-time, meaning that orders arrive instantaneously once they are placed. However, the full derivation of the EOQ with back-ordering model is outside the scope of this text.

TIC = Total Ordering Cost + Total Carrying Cost + Total Shortage Cost

$$= \text{TOC} + \text{TCC} + \text{TSC} = \frac{F \times S}{Q} + \frac{M^2 \times C \times P}{2Q} + \frac{(Q - M)^2}{2Q} s' \qquad (13\text{-}22)$$

where F = fixed ordering cost; S = annual demand/units sold; Q = ordering quantity; M = maximum inventory; C = carrying cost percentage; P = price per unit; and s' = shortage costs.

With back-ordering, all the components of cost need to be combined to get the TIC. These include total ordering costs, total carrying costs, total shortage costs, total unit costs, and total back-ordering costs (Waters 2003). It is important to consider the sum of these costs because taking its partial derivative with respect to Q yields the optimum value for order size. The TIC in this case is transformed to the total inventory cost per cycle.

In Eq. (13-23), we take the partial derivative of the total inventory costs with respect to Q to get the optimum value for the order size, Q^* (Gupta and Khanna 2011; Waters 2003).

$$\frac{\partial(\text{TIC})}{\partial Q} = \frac{F \times S}{Q} + \frac{M^2 \times C \times P}{2Q} + \frac{(Q - M)^2}{2Q} s' = 0 \qquad (13\text{-}23)$$

After the partial derivation, we find that the total inventory cost is minimized for Q^*, where Q^* is the optimal order quantity. Solving equation (13-23) and setting it equal to zero, along with some manipulation, we get Eq. (13-24) (Gupta and Khanna 2011; Tersine 1994):

$$Q^* = \left[\frac{2FS(CP + s')}{CPs'}\right]^{0.5} \qquad (13\text{-}24)$$

In Eq (13-25). the total back-order quantity, B, is derived based on Q^* multiplied by a factor including carrying and shortage costs per unit. Hence,

$$B = Q^* \left(\frac{CP}{CP + s'} \right) \tag{13-25}$$

Now we can use the above equations in an example.

13.10.1 Back-Ordering Example

SpeedRun has decided to explore the costs associated with allowing back orders. Management has determined that the company will incur a cost of $18 for each pair of shoes that it is short each year. Annual demand remains at 13,000 pairs, the fixed ordering cost at $500 per order, the carrying cost at 20%, and the purchase price at $28.89. Allowing for back orders, what are the optimal order quantity, maximum inventory level, and maximum shortage that SpeedRun will experience? Assume a zero lead time.

To start, we know that

$$S = 13,000 \, \text{pairs} \quad F = \$500 \quad C = 0.20 \quad P = \$28.89 \quad s' = \frac{\$18}{\text{pair}}/\text{year}$$

Thus, we can solve for Q^* by substituting the above values into the respective equations for optimal order quantity and maximum inventory level:

$$Q^* = \left[\frac{2FS(CP + s')}{CPs'} \right]^{0.5} = \left[\frac{2(\$500)(13,000)[(0.2)(\$28.89) + \$18)]}{(0.2)(\$28.89)(\$18)} \right]^{0.5} = 1,724 \, \text{pairs}$$

The total back-order quantity can now be obtained by substituting the value for Q^* and the other given variables into the equation for total back-order quantity:

$$B = Q^* \left(\frac{CP}{CP + s'} \right) = 1,724 \left(\frac{(0.2)(\$28.89)}{(0.2)(\$28.89) + (\$18)} \right) = 1,724(0.243) = 419 \, \text{pairs}$$

Thus, we have determined that, when back orders are allowed, the optimal order quantity is 1,724 pairs; the optimal maximum inventory is 1,305 pairs; and the back-order quantity is 419 pairs.

13.11 Ordering with Quantity Discounts

In real-world practice, a company often has the option of placing a larger order in exchange for a discount on the purchase price per item. This may be beneficial, depending on the magnitude of the discount and the holding cost per item. Though carrying costs would increase because inventory held would increase, placing a larger order means that fewer orders must be placed to meet annual demand, which in turn reduces annual ordering costs. If the quantity discount and order cost savings do not outweigh the holding cost associated with taking on the extra stock, it would be uneconomical to place a larger order. We can illustrate this with another example from SpeedRun.

13.11.1 Quantity Discounts Example

Suppose that SpeedRun is offered a quantity discount of 5% on orders of 2,000 pairs of shoes or more. Applying this quantity discount to the purchase price, we have a discount price of $(1 - 0.05) \times (\$28.89) = \$27.446$. Since the previously determined EOQ is 1,500 pairs, SpeedRun will have to place an order greater than the EOQ to obtain the quantity discount. In order to determine whether or not to place a larger order to receive the discount, despite the increase in inventory carrying costs, we will compare the total inventory costs of ordering at EOQ and at the minimum for a discount, 2,000 pairs. As we have calculated earlier, the total inventory cost when ordering at EOQ is in Eq. (13-26):

$$\text{TIC} = \frac{F \times S}{Q} + \left(C \times P \times \frac{Q}{2} \right) = \frac{(\$500)(13,000)}{1,500} + (0.2)(\$28.89)\frac{1,500}{2}$$
$$= \$4,333.33 + \$4,333.33 = \$8,666.83 \qquad (13\text{-}26)$$

We now calculate the total inventory cost for the case of ordering 2,000 pairs of shoes instead of the EOQ of 1,500. The only terms to change will be P and Q. Therefore, Eq. (13-27) gives us

$$\text{TIC} = \frac{(\$500)(13,000)}{(2,000)} + (0.20)(\$27.446)\frac{2,000}{2}$$
$$= \$3,250.00 + \$5,489.20 = \$8,739.20 \qquad (13\text{-}27)$$

Examining these two cases reveals a few differences. In the discount case, ordering costs decrease because of fewer orders being placed in a year, while holding costs increase because of placing an order large enough to acquire the discount. Second, the purchase price, P, is 5% less in the discount case. Comparing the total inventory costs for both cases, we see that the company would actually spend $72.37 more per year by placing orders large enough to acquire the quantity discount, even though it appears, up front, that the price is lower by 5%. Total inventory cost is actually larger in the quantity discount case, which leads us to the conclusion that we should continue to order at EOQ to incur the least cost. Although the discount may save on the individual unit purchase price, this savings is less than the increase in holding cost owing to the larger quantity that we must hold in inventory (Plossl 1985). In other situations, it may be advisable to take the discount; each individual case must be evaluated in itself.

Even though we have determined that it is not reasonable to place a larger order to obtain the 5% discount, we may want to explore what minimum discount level would make the larger order economical. This is important because the minimum discount level can be used by SpeedRun to negotiate a deal with the manufacturer that will benefit the company. To determine this percentage discount, we must equate the total inventory cost from the quantity discount case to the total inventory value from the EOQ case, leaving the purchase price as a variable, P. We then solve for P.

$$\text{TIC} = \frac{(F)(S)}{Q} + (C)(P)\frac{Q}{2}$$

Therefore,

$$\$8,666.83 = \frac{(\$500)(13,000)}{2,000} + (0.20)(P)\left(\frac{2,000}{2}\right) = 3,250 + 200 \times P$$

Yielding $P = \$27.084$. (13-28)

From which the discount is calculated in Eq. (13-29):

$$\text{Discount}_{min} = 1 - \frac{\$27.084}{\$28.89} = 0.0625 = 6.25\% \tag{13-29}$$

The minimum discount is found to be 6.25% for a minimum order size of 2,000 pairs of shoes. In other words, it will only be economical for SpeedRun to purchase a quantity of 2,000 pairs of shoes if the discount on the purchase price is greater than 6.25%. Purchasing a 2,000-pair quantity at any discount lower than 6.25% will incur excessive costs to the company because of the uneven tradeoff between holding costs and ordering costs.

13.12 Exercises

Use the following information for questions (a) through (d):

Red Dirt Inc. is a contractor that specializes in mass grading on construction projects. They have a fleet of heavy equipment, ranging from scrapers to dozers, that is projected to use a total of 50,000 gallons of diesel throughout the year. The cost of diesel to Red Dirt is $4.16 per gallon; there is a 15% cost of carrying inventory; and the cost of placing an individual order is $100.

   (a) What is the optimal quantity of diesel fuel that should be ordered? How often should orders be placed? Assuming a three-day lead time, what is the reorder point?

Table 13-3. Probability Distribution for Diesel Consumption in an Order Period

Probability	Diesel consumption (gallons)
0.125	3,000
0.25	3,500
0.25	4,000
0.25	4,500
0.125	5,000
$\sum$ prob. = 1.0	Expected sales = 4,000

(b) Red Dirt Inc. wishes to determine how much safety stock should be carried. Because of the nature of the company's work, the shortage cost is $50 for each gallon of diesel that is short. This shortage cost includes loss of productivity, overtime costs, increased labor costs, and so forth. Using the probability distribution of diesel usage in Table 13-3, perform a safety stock analysis and recommend the appropriate amount of safety stock that should be carried.

(c) Determine the maximum inventory level, the optimal order size, and the maximum shortage that Red Dirt Inc. should expect if it allows back orders. Assume a shortage cost of $50.

(d) The diesel supplier has approached Red Dirt Inc. with a proposal of a 6% quantity discount on orders of 5,000 gallons or greater. Should Red Dirt Inc. accept the quantity discount offer and order 5,000 gallons of diesel at a time?

References

Brigham, E. F., and Gapenski, L. C. (1985). *Financial management: Theory and practice*, CBS College, New York.

Gupta, M. P., and Khanna, R. B. (2011). *Quantitative techniques for decision making*, 4th Ed., PHI Learning Private Limited, New Delhi, India.

Plossl, G. W. (1985). *Production and inventory control: Principles and techniques*, Prentice-Hall, Englewood Cliffs, NJ.

Tersine, R. J. (1994). *Principles of inventory and materials management*, PTR Prentice-Hall, Englewood Cliffs, NJ.

Thomas, A. B. (1969). *Inventory control in production and manufacturing*, Cahners, Boston, MA.

Toomey, J. W. (2000). *Inventory management: Principles, concepts and techniques*, Kluwer Academic, Boston.

Waters, C. D. J. (2003). *Inventory control and management*, Wiley, Hoboken, NJ.

Winston, W. L. (1991). *Operations research applications and algorithms*, 2nd Ed., PWS-Kent, Boston.

Case Study for a Risk Plan: Kapi'olani Boulevard

A1.1 Introduction

Risk is the possibility that an action will affect a given project or event in any field. So risk management is the study of the process and structure that manage the adverse effects of risk. Hence, risk management aims not to eliminate all risks from a given project, but rather to recognize that risks challenge a project and to develop an appropriate strategy to deal with them (Zou et al. 2010). Very often, risks can be turned into opportunities that were hitherto unknown; at other times, one can mitigate the risks just by knowing they exist.

A1.2 Background

A successful project consists of many aspects such as scope, design, cost, and schedule management. A sometimes overlooked component of projects is risk management. It is usually not a standard part of project development or construction management, which often focus on cost, schedule, and quality. However, implementing risk management in construction projects may greatly benefit them and should therefore be considered an essential part of such projects (FHWA 2006).

Today, more and more megaprojects are being planned and constructed throughout the world. Huge in scope, investment, and performance, megaprojects have experienced some of the best and worst planning, engineering, and construction. Infrastructure megaprojects provide connectivity and mobility as never seen before but have also resulted in huge cost overruns, delays, and technical failures. Especially for megaprojects, a rigorous, detailed, and comprehensive risk management system is necessary (Flyvbjerg 2008). History shows that megaprojects tend to encounter severe cost overruns. For example, the cost overrun for the Channel Tunnel between England and France was 80%, for Denver's new international airport it was close to 200%, and for the Great Belt Link in Denmark it was 54%. Both ordinary and megaprojects encounter these problems because risk is not addressed at all. Instead, projects are planned using an "everything goes according to plan" principle, which does not consider possible risks or their effects on a project. Instead, projects should be planned and assessed based on a "most likely development" principle, where risk identification, assessment, and analysis of risks are incorporated in developing the project, and a comprehensive risk management plan is developed to address those risks (Flyvbjerg et al. 2003).

A1.3 Risk Management

The six main steps in risk management are (1) *identification,* (2) *assessment,* (3) *analysis,* (4) *mitigation,* (5) *allocation, and* (6) *monitoring and updating.* Risk management is a repetitive and cyclical process that continues throughout the design and construction of a project: As a project evolves, new risks may develop, while others are resolved. (See Fig. A1-1 for the risk management flow chart.) The following sections will describe all six steps of risk management in detail using a fictional example, a roundabout at Kapiʻolani Boulevard and Ward Avenue in Honolulu, for illustration.

A1.3.1 Roundabout at Kapiʻolani Boulevard and Ward Avenue

Honolulu's City Department of Traffic Services (CTDS) decides to develop a risk management and allocation program. Honolulu, like most cities across the United States, has to deal with aging infrastructure, tight funding, and a backlog of

Fig. A1-1. Risk management flow chart

maintenance. CTDS implemented the new risk management and allocation program because of a history of severe cost escalation of projects, legal actions from stakeholders, and construction management problems and mistakes. The department believes that this new risk management system will improve all stages of project management, from planning to engineering to construction. The first project to implement this system is the construction of a roundabout at the intersection of Ward Avenue and Kapi'olani Boulevard.

Project Description: CTDS has decided to design and build a roundabout to address congestion and safety concerns in the existing at-grade intersection of Ward Avenue and Kapi'olani Boulevard, as shown in Fig. A1-2. This is an important intersection in Honolulu.

Fig. A1-2. Map of Kapi'olani Boulevard and Ward Avenue intersection (from Google Maps)

The location of the star represents the proposed location of the roundabout, the exact size of which has not yet been determined; numerous political and legal steps will need to be taken, possibly including the exercise of eminent domain to claim the land and properties around that intersection to construct the roundabout. The project is considered of average size and technical complexity, is in the preliminary stages, and has the following scope and characteristics:

- The project will convert the Ward Avenue and Kapiʻolani Boulevard intersection into a roundabout; traffic signals will be removed.
- Ward Avenue currently consists of six 12-foot lanes with no shoulder.
- Ward Avenue has dedicated left-turn and right-turn lanes in both the northbound and southbound directions.
- Kapiʻolani Boulevard currently consists of six 12-foot lanes with no shoulder.
- Kapiʻolani Boulevard has dedicated left-turn lanes in both the eastbound and westbound directions.
- Right of way is available on the north side of Kapiʻolani Boulevard.
- Right of way is also available on the west side of Ward Avenue.

A1.4 Risk Identification

The first step in risk management is to identify and categorize the possible risks that could affect a project and then to document them.

The risk identification process depends on the project. For small projects, risk identification can mean a list of issues that can be assigned to individual project members who will work on them throughout the project. However, this strategy is not sufficient for larger and more complex projects. In this case, the identified risks have to go through the entire process explained in the following sections.

Risk identification can be based on historical data, checklists, and team knowledge and should allow the project manager to understand the risks and their causes. After the risks are identified, they should be divided into subgroups of similar risks. This classification of risks prevents redundancy and makes the following stages of risk management easier (FHWA 2006).

A1.4.1 Risk Identification for the Roundabout at Kapiʻolani Boulevard and Ward Avenue

Because this project will be the first to implement the new risk management and allocation system, CTDS conducted a facilitated workshop to help identify risks. The project team collected data such as project description, cost estimates and design, and construction schedules. The facilitator brought standardized risk checklists to ensure that the project team did not leave out any possible risks. The workshop began with a brainstorming session in which the team members, based on their knowledge and experience, made a list of possible risks. The risks were then divided into several subgroups:

Technical Risks:

- The right-of-way analysis is in error at Ward Avenue south of Kapiʻolani Boulevard.
- Inaccurate assumptions about technical issues were made in the planning stage.

External Risks:

- Property owners are unwilling to sell land at the intersection.
- Local communities are concerned about traffic flow during construction.

Environmental Risks:

- The area contains a historic site, possibly including native graves.
- The ground is polluted.

Project Management Risks:

- The project scope, schedule, and costs are not clearly described.
- There is pressure to deliver the project on an accelerated schedule.

A1.5 Risk Assessment

After the risk identification comes the risk assessment. A risk assessment consists of evaluating the identified risks and has two steps: determining the probability of the risk occurring and analyzing the impact if the risk occurs.

A comprehensive risk assessment combines qualitative and quantitative assessments. The qualitative assessment evaluates risk based on a worst-case scenario and its likelihood. The quantitative assessment evaluates risk in terms of tangible results and its likelihood. The qualitative assessment effectively screens and prioritizes the risks and allows the development of appropriate mitigation and allocation strategies; the quantitative assessment determines the numerical and statistical nature of the risks.

The risk assessment must identify who owns the risk. The owner carries some risk; the project developer or contractor carries other risks; and the stakeholders share some risk.

As stated previously, the two steps of risk assessment are determining the probability that a given risk will happen and evaluating the risk's potential impact on the project. One method of classifying risks is to develop a two-dimensional matrix that divides risks into three categories based on their frequency and severity.

Table A1-1. Determination of the Likelihood of Risk

Level	Likelihood
A	Remote
B	Unlikely
C	Likely
D	Very likely
E	Near certain

For the first category, we determine the *likelihood* that a risk will happen. Five stages of likelihood were used, designated A through E and ranging from remote (A) to near certain (E), as shown in Table A1-1. A larger or smaller likelihood scale may be used if appropriate—a seven-point scale from A to G or a three-point scale from A to C, for example.

For the second category, we determine the *consequences* of a given risk if it happens. Again, five levels of consequences or impacts are used, ranging from minimal (a) to unacceptable (e), as shown in Table A1-2.

For the third category, we combine the likelihood and consequences into one matrix, which classifies the risk as low (L), moderate (M), or high (H), depending on its frequency and severity. The classification is done by addressing both the likelihood and the consequences of each risk. For example, a risk with a remote likelihood of occurrence and minor consequences will be classified as low risk, and a risk deemed highly likely with moderate severity will be classified as a moderate risk. The Ls, Ms, and Hs are shown with different shading in Table A1-3.

Low-risk events will cause minimal impact and need minimum oversight to keep the risk low. Moderate-risk events will moderately disrupt the project; dealing with the risk at hand may require a different approach, and the project manager will have to provide additional oversight. High-risk events will significantly disrupt a project, and tackling them will require a different approach. Thus, project managers must prioritize their risks.

In the analysis of risks, low risks can usually be discarded and eliminated from further assessment. Moderate risks either have a high likelihood with minor consequences or a low likelihood with major consequences. Risks with high likelihood

Table A1-2. Determination of the Consequences of Risk

Level	Consequences
a	Minimal
b	Minor
c	Moderate
d	Major
e	Unacceptable

Table A1-3. Classification Matrix

Likelihood		a	b	c	d	e	Importance Scale
	E	M	M	H	H	H	**Importance Scale**
	D	M	M	M	H	H	L = Low risk
	C	L	L	M	M	H	M = Moderate risk
	B	L	L	L	M	M	H = High risk
	A	L	L	L	L	M	
		a	b	c	d	e	

Consequences

and minor consequences will usually affect projects minimally when taken individually. However, the aggregate and cumulative effects of several risks of high likelihood with minor consequences can have a significant impact on the project, so they must still be analyzed, mainly because their tendency to occur frequently can continuously nag at the project team. Risks with a low likelihood and major consequences should also be analyzed further, because they can potentially have a cumulative impact on the project. High risks naturally require further analysis (FHWA 2006).

A1.5.1 Risk Assessment for the Roundabout at Kapi'olani Boulevard and Ward Avenue

The CTDS team assessing the risks for this project categorized all the identified risks by determining their likelihood and consequences. The team members' experience and knowledge from previous projects contributed to this determination. The frequency and severity of the identified risks were quantified and ranked. Risks found to have a low impact on the project were not subjected to further analysis. Risks found to have a moderate impact, either because of high likelihood with minor consequences or low likelihood with major consequences, were subject to further analysis, as were risks found to have a high impact. Tables A1-4 to A1-6 show the assessment of the three risks determined to have a high impact, all falling in the dark cells of the tables.

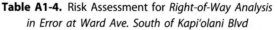

Table A1-4. Risk Assessment for *Right-of-Way Analysis in Error at Ward Ave. South of Kapi'olani Blvd*

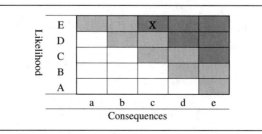

Table A1-5. Risk Assessment for *Historic Site, Possibility of Native Graves in Area*

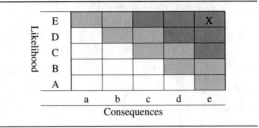

Table A1-6. Risk Assessment for *Pressure to Deliver Project on Accelerated Schedule*

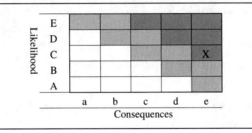

Table A1-4 shows the assessment for the risk that the right-of-way analysis is in error at Ward Avenue south of Kapi'olani Boulevard. This risk was assessed to have a *near certain* likelihood and a *moderate* impact on project completion. Table A1-5 shows the risk of historic site, possibly native graves in area was assessed as having a *near certain* likelihood and *unacceptable* consequences for the project. Lastly, Table A1-6 shows that pressure to deliver the project on an accelerated schedule was assessed to have a *likely* probability and *unacceptable* consequences for the project.

A1.6 Risk Analysis

After the risk assessment, the risks deemed to need further analysis are analyzed. Risk analysis aims to combine the identified and assessed risks into an overall project risk estimate. Project owners can determine whether a project should progress, and, if so, to determine cost and schedule contingency values.

Several methods are available for conducting a risk analysis. Each of the methods has a trade-off between sophistication and ease of use: The more sophisticated the method, the more difficult it is to use, and vice versa. The output depends on the choice of method; it can be either a single parameter or the complete distribution. The four most common methods used for risk analysis are the traditional method, the analytical method, simulation models, and probability trees.

In the *traditional method*, each risk is assigned a risk factor, usually based on historical knowledge, and the classification determined in the risk assessment step. Hence, a risk that is classified as high risk will have a greater risk factor than a risk that is classified as moderate risk. The project contingency is then found by multiplying the risk factor by the estimated cost of each element. In the example project, the traditional method will be used.

The *analytical method* uses estimates of both the mean and the standard deviation to provide a contingency estimate. In this method, the variance in the output is shown, and the estimated mean and standard deviation are used as risk factors to determine contingency funds. The greater the standard deviation of the data, the greater the possible fluctuation and chance of exceeding safety levels. The Bayesian method of analyzing events for informed decision making was presented in Chapter 8. There are also other statistical methods and tests that can help analyze quantitative risk, but they will not be covered in this text.

Simulation models are computer-based probabilistic calculations that draw samples from probability distributions. They are relatively more complex to use but provide more detailed information about risk impacts on cost and schedule. However, the additional information gained does not always justify the time and cost of the simulation techniques.

Finally, *probability trees* are simple diagrams that show the impact of sequential events. This method is effective for determining the interrelationship between risks (FHWA 2006).

These methods can be as complex and sophisticated as desired, but their time and cost may not always be justified. In contrast, the traditional method is easy to implement and works effectively for decision-making purposes.

A1.6.1 Risk Analysis for the Roundabout at Kapi'olani Boulevard and Ward Avenue

The risk assessment showed the need to conduct a rigorous analysis of the risks deemed high so as to develop a comprehensive risk management plan and to determine the necessary contingency funds. The project team used the traditional method to calculate the contingency funds for the risks classified as high risk. This analysis considered only the following three risks, which were found to have the highest risks in the project:

- Risk 1: The right-of-way analysis is in error at Ward Avenue south of Kapi'olani Boulevard.
- Risk 2: The area contains a historic site, possibly including native graves.
- Risk 3: There is pressure to deliver the project on an accelerated schedule.

The risk factors used to determine the contingency funds for this project were based on a combination of historical knowledge, team members' experience, and the risk assessment. Risk 1 has a near certain likelihood of occurrence but only moderate consequences, thus being assigned a risk factor of 20%. Risk 2 has a near certain likelihood of occurrence and unacceptable consequences, which the team members decided gave it a risk factor of 30%. Similarly, using their wealth of knowledge and experience, the team members assigned a risk factor of 10% to Risk 3.

Contingency funds are found by multiplying the risk factor, which is a percentage value, by the estimated cost of the risk. Table A1-7 shows the contingency funds necessary to deal with the project's risks. (Chapter 3 also presented a technique for determining contingency funds.)

A1.7 Risk Mitigation and Planning

Once risks have been identified, assessed, and analyzed, the involved parties are in a better position to determine the best course of action to mitigate those risks. Remember that the objective of risk management is not necessarily to eliminate risks, but to recognize that risks do exist and to manage them. Questions before a risk management strategy is devised include

- What can be done?
- What options are available?
- What are the costs and benefits of these options? and
- What are the impacts of current decisions on future events?

Table A1-7. Contingency Funds

Risk	Estimated cost	Risk factor (%)	Contingency funds
1: Right-of-way analysis in error at Ward Ave. south of Kapiʻolani Blvd.	$380,000	20	$76,000
2: Historic site, possibility of native graves in area	$100,000	30	$30,000
3: Pressure to deliver project on accelerated schedule	$250,000	10	$25,000

Possible ways of dealing with risk are ignoring the risk, recognizing the risk but not responding to it, avoiding the risk by dealing with it, reducing the risk through mitigation, transferring the risk to others through contract or insurance, or retaining and absorbing the risk. Developing a risk management strategy requires different levels of detail depending on the size and complexity of the project.

The option of identifying issues to track possible risks was mentioned previously. More complex projects require risk charters to help manage the risks. A risk charter describes the identified risks with estimates of their cost and scheduling impact. A risk charter can take many forms, but usually includes risk description, response strategy, status, date identified, project phase, functional assignment, risk trigger, probability of occurrence, impact, response actions, and responsibility (FHWA 2006).

To better track and manage risks, CTDS developed a basic risk charter. Considerable work must be done to prepare this risk charter, which requires a dedicated team of professionals. This risk charter includes practical action items: risk description, response strategy, status, project phase, and responsibility. Table A1-8 shows the charter for the three high-risk events.

A1.8 Risk Allocation

The project contract defines the basis for risk allocation and the roles and responsibilities of all parties involved; it also covers cost, time, quality, delays, disputes, and claims. When risks are analyzed and their consequences identified and measured, decisions about allocating risks in a way that minimizes cost and promotes project goals can be made. Any party assuming risk must be prepared for the financial burden associated with that risk. The contingency funds necessary for each risk have been determined during the risk analysis.

In some cases, multiple parties can share the risk. For example, a commonly shared risk is unusual weather conditions. Often the contractor receives a time extension because of severe weather conditions but does not receive financial compensation for the extra cost. With this technique, the contractor and the owner share the risk of severe weather conditions.

A common tool for risk allocation is compiling the risks in a table or matrix. For some projects, a combined risk mitigation and planning and risk allocation table is created because similar things are included in both mitigation and allocation, such as risk description, strategy, status, project phase, action, and who is responsible. The table is used to allocate the risks in the contract and can serve as a communication tool during the design and construction phases (FHWA 2006).

To keep the analysis simple, Table A1-8 doubles as the risk allocation table. This table comprises a description of each high-risk event, response strategy and action, status, project phase, and the person responsible for the management of the given risk. Such a strategy and table is essential to ensuring that risks are addressed and tackled.

Table A1-8. Risk Charter

Risk	Response strategy	Response action	Status	Project phase	Responsibility
1: Right-of-way analysis in error at Ward Ave. south of Kapiʻolani Blvd.	Avoidance	The team will attempt to design around the areas where right of way may be an issue.	Completed	Design phase	Design team lead
2: Historic site, possibility of native graves in area	Acceptance	This issue will be dealt with according to local laws and regulations if it occurs.	Occurred	Construction phase	Construction lead
3: Pressure to deliver project on accelerated schedule	Mitigation	The team will follow time schedules strictly. There will be weekly follow-up to ensure that the project is on time.	Did not occur	Construction phase	Construction lead

A1.9 Risk Monitoring and Updating

Risk monitoring and updating aim to systematically track the identified risks, identify any new risks, and effectively manage the contingency funds. This serves as the final step in risk management. However, it must be continued throughout the life of the project because risks are dynamic: They will most likely change over the course of the project, necessitating monitoring and updating. As explained previously, the risk management system is a circular and repetitive process; therefore, the previously described steps must be repeated if new risks are identified during the process (see Fig. A1-1).

Carefully documenting the risk management process is important because it contributes to program assessment and updates as projects develop. In addition, formal documentation tends to ensure more comprehensive risk analysis, thus providing a basis for monitoring mitigation and allocation actions, which allows it to provide background information for new personnel and project decisions (FHWA 2006).

A1.9.1 Risk Monitoring and Updating for the Roundabout at Kapi'olani Boulevard and Ward Avenue

As a part of the CTDS risk management plan, a strict monitoring and updating scheme was implemented. CTDS decided to conduct weekly project risk reviews in case any risks developed or new, unforeseen risks emerged. CTDS developed a project risk-status report (shown in Table A1-9), showing all risks, their impacts (high, moderate, or low) and the current status of each risk. The table includes both expected and newly identified risks.

Table A1-9 indicates that three new risks—4, 5, and 6—were identified and that two of these risks (unidentified utilities on Kapi'olani Boulevard and breakage of an unidentified water pipe on Kapi'olani Boulevard) affected the already high-risk issue of finishing the project on time. Thus, the project is now behind schedule.

The engineers can take it from here and see that the cycle of risk management repeats after monitoring and updating per the risk management cycle of Figure A1-1. New risks must be identified and the entire process must be taken up again.

A1.10 General Discussion

This appendix explained why risk management should be considered an essential part of project development and construction. Risk management does not aim to eliminate risks but rather attempts to manage risks so as to minimize costs and time overruns and optimize the project's goals. Small, medium, and megaprojects can benefit from developing a risk management strategy, the form and size of which

Table A1-9. Risk-Status Report

Risk	Impact			Comment/status
	High	Moderate	Low	
1: Right-of-way analysis in error at Ward Ave. south of Kapiʻolani Boulevard.		✓		Closed; design team designed around affected areas
2: Historic site, possibility of native graves in area	✓			Most digging completed
3: Pressure to deliver project on accelerated schedule			✓	Behind schedule due to discovery of unexpected utilities
4: Local communities concerned with traffic flow during construction		✓		Continued concern with maintenance of traffic flow
5: Breakage of unidentified water pipe on Kapiʻolani Boulevard			✓	Closed
6: Unidentified utilities discovered on Kapiʻolani Boulevard		✓		Utilities identified; contractor will move affected utilities

depends on the project. The strategy can include lists of possible issues, risk charters, strategies, actions, persons responsible, and follow-up procedures.

Risk management is a circular process, and it must be conducted throughout the entire project process to yield the best result. Because projects are dynamic and events interact with one another, continuous risk management is necessary.

A1.11 Case-Study Problem

Hilo, Hawaii, has long struggled with street flooding during heavy rains. The county sewer department has therefore decided to upgrade the sewer system on Kanoelehua Avenue between Kamehameha Avenue and Puainako Street as a first

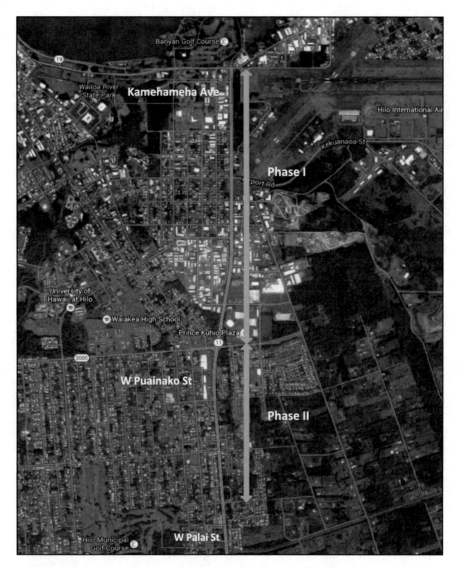

Fig. A1-3. Map of project area in Hilo, HI (from Google Maps)

phase, and between Puainako and Palai Streets as a second phase. Fig. A1-3 shows the project location and size.

This project is of average size and complexity. It is in the preliminary stages and has the following scope and characteristics:

- In Phase I, the project will upgrade the sewer system on Kanoelehua Avenue between Kamehameha Avenue and Puainako Street.
- In Phase II, the project will upgrade the sewer system on Kanoelehua Avenue between Puainako Street and Palai Street.

- Sufficient right of way is available during Phase I.
- Limited right of way is available during Phase II.
- The project is based on old utility plans.

Make other assumptions as necessary. Then develop a risk management plan following the six steps explained in this chapter, and discuss the following issues:

- What are the benefits of a good risk management system?
- Why is risk management a circular process?
- What tools can you use to identify possible risks?
- What is the difference between qualitative and quantitative risk analysis?
- What are some different responses to or actions regarding risk?
- What document allocates risk?

References

FHWA (Federal Highway Administration) (2006). "Risk assessment and allocation for highway construction management." Rep. No. FHWA-PL-06-032, Office of International Programs, Office of Policy, Washington, DC.

Flyvbjerg, B. (2008). *Public planning of mega-projects: Overestimation of demand and underestimation of costs*, Edward Elgar, Northampton, MA.

Flyvbjerg, B., Bruzelius, N., and Rothengatter, W. (2003). *Megaprojects and risk: An anatomy of ambition*, Cambridge University Press, Cambridge, U.K.

Zou, P. X. W., Chen, Y., and Chan, T.-Y. (2010). "Understanding and improving your risk management capability: Assessment model for construction organizations." *J. Constr. Eng. Manage.*, 10.1061/(ASCE)CO.1943-7862.0000175, 854–863.

Example Contingency Fund Evaluation Problem

This appendix is an extension of Chapter 3 and presents a direct, mechanical process for evaluating contingency funds.

A new high-class apartment building will be constructed in a downtown area. The client is trying to procure a well-suited contractor. Bidding is now open, and contractors have picked up the bid documents. As a contractor, you would like to place a bid but are unsure about the proper amount to use in estimating contingency. The procedure for outlining the process and solving this problem for the four hypothetical work packages in Table A2-1 is as follows:

1. Establish critical and noncritical items based on the 0.5% criterion.
2. Use Monte Carlo to generate 10 sets of random numbers.
3. Draw an overrun profile.
4. Draw a contingency drawdown curve showing percentages of contingency funds consumed versus time elapsed and contingency fund versus time.

A2.1 Steps to Solution

I developed this step-wise solution while teaching graduate students and found that all of them easily grasped the procedure.

Step 1: Check whether all activities satisfy the 0.5% criticality rule

The total target cost for the four work packages is $30 million.

- 0.5% of $30 million is $0.15 million.
- But every work package fluctuates—by at least $1 million—between its low price and target price *or* its high price and target price.
- $1 million is greater than $0.15 million.
- Hence, all work packages are critical.

Table A2-1. Costs for Apartment Building

Work packages	Low ($ million)	Target ($ million)	High ($ million)	Probability	Project progress when completed (% complete)
Utilities	5	6	7	Low = 0.2 Target = 0.4 High = 0.4	25
Concrete structure	11	12	15	Low = 0.2 Target = 0.4 High = 0.4	45
Steel	7	8	10	Low = 0.2 Target = 0.3 High = 0.5	80
Interior works	3	4	7	Low = 0.2 Target = 0.2 High = 0.6	100

Step 2: Set up the Monte Carlo

In setting up the Monte Carlo slab for generating random numbers, the low, target, and high cost probabilities given in Table A2-1 are used. For utilities, the chance of getting a low price of $5 million is 20%. Hence, the Monte Carlo slab for the low price must span a 20% band that can be represented by the generation of a random number between 0.0 and 0.2. Similarly, the chance of getting a target price of $6 million, which is higher than the low price, is 40%. Now the Monte Carlo slab for the target price must generate a random number between 0.2 and 0.6. Finally, the high price of $7 million can be represented by the remaining probability, 0.6 to 1.0. The Monte Carlo slabs for all work packages are shown in Table A2-2.

Step 3: Generate random numbers and record values

Random numbers for selecting the appropriate price of each work package in any simulation, per the Monte Carlo slab, can be generated with an ordinary scientific calculator or in Excel by using the RANDOM function. These random numbers are invariably between 0 and 1, because probabilities can only range from 0 (impossible) to 1 (certain). Suppose now that for the utilities work package, the first random number generated is 0.8475. This number corresponds to the high price of $7 million. Hence, $7 million will be recorded as the price for utilities in simulation run I (see Table A2-3). Similarly, for the first simulation run we get prices of $15 million, $10 million, and $7 million for concrete structures, steel, and interior works, respectively. These numbers came about because we got the random

Table A2-2. Monte Carlo Slabs

	Value	Monte Carlo slab
Utilities	5	0–0.2
	6	0.2–0.6
	7	0.6–1.0
Concrete structure	11	0–0.2
	12	0.2–0.6
	15	0.6–1.0
Steel	7	0–0.2
	8	0.2–0.5
	10	0.5–1.0
Interior works	3	0–0.2
	4	0.2–0.4
	7	0.4–1.0

Table A2-3. Costs Based on Random Number Generation

	I	II	III	IV	V	VI	VII	VIII	IX	X
Utilities	7	7	5	7	7	5	6	6	7	7
Concrete structure	15	15	11	12	15	11	12	12	15	12
Steel	10	10	7	8	10	7	10	10	10	8
Interior works	7	7	3	4	7	3	7	7	7	4
Total	39	39	26	31	39	26	35	35	39	31

numbers 0.7244 for concrete structures, 0.6215 for steel, and 0.5795 for interior works, resulting in a total price of $39 million for the first simulation run.

Likewise, the prices for each work package in the other simulation runs are recorded. For simplicity, only ten simulation runs were undertaken for this example; generally, the greater the number of simulation runs, the better the accuracy of the results.

Step 4: Determine the frequency distribution for the total cost

For the next step, arrange the total costs for the simulation runs from Table A2-3 in ascending order and count the number of times each total price occurs. This will be the basis of the frequency distribution we need. Ignore all noncritical items because they are not operational in risk analysis and will not contribute to the contingency fund. So $26 million, $31 million, and $35 million each occur two times, and $39 million occurs four times. This frequency distribution represents the probability of the actual distribution in reality. The cumulative probability must sum to 1.0, which it does for the cost of $39 million. Table A2-4 shows the data for the frequency distribution and cumulative probability.

Table A2-4. Probability Distribution and Cumulative Probability

Value ($ million)	Non-critical ($ million)	Total price ($ million)	Frequency count	Probability	Cumulative probability
26	0	26	2	1/5	1/5
31	0	31	2	1/5	2/5
35	0	35	2	1/5	3/5
39	0	39	4	2/5	5/5 = 1

Step 5: Draw the cumulative probability profile

Drawing the cumulative probability profile in Fig. A2-1 is a mere stone's throw from the data in Table A2-4. The tabular data is simply transferred to a graphical format.

Step 6: Determine the acceptable confidence level and calculate the contingency fund

The contractor must ascertain at what confidence level he or she wishes to operate. Each person has a different aversion to or propensity for risk; hence, the confidence level is a function of how much stomach one has. There are dangers in being too risky (low confidence level), but there are other types of risks in being too conservative (high confidence level). Assume that the contractor settles on

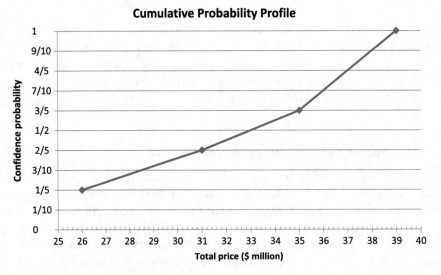

Fig. A2-1. Cumulative probability profile

a confidence level of 80%, for which the corresponding bid price from Fig. A2-1 is $37 million.

Hence, the contingency fund, which is the bid price minus the target price, will be $37 million minus $30 million, or $7 million.

Step 7: Distribute the contingency fund by calculating the expected overrun

To distribute the contingency fund of $7 million, one must know how to allocate it to the various work packages. First of all, there is no need to allocate funds to a work package that has a net expected value of underrun rather than overrun because there is a limited chance of risk in such a scenario. Thus, contingency funds must only be allocated to those work packages that show a potential expected net overrun in the balance. Multiplying the probability of underrun by the dollar amount of underrun from the target price gives us the expected underrun. Similarly, we can find the expected overrun by multiplying the probability of overrun by the dollar amount of overrun. The difference between expected overrun and expected underrun will determine whether there is a net underrun or a net overrun.

Adding all expected overruns gives the total expected overrun. It follows that each work package with an overrun will collect contingency funds based on the proportion of its expected overrun to the total expected overrun. These data are all shown in Table A2-5. It turns out—using calculations that are self-explanatory—that the percentage share of the contingency funds for utilities is 5.56%, for concrete structure is 27.78%, and so on.

Multiplying the percentage share (percentage allocation) by the total contingency amount yields the final contingency fund allocation for each work package. As a check of the calculations, the total contingency fund allocation should add up to the contingency fund amount—in this case, $7 million.

Step 8: Plot the drawdown curve

The drawdown curve is simply the speed or rate at which the contingency funds should be depleted, or consumed. This is an important guide for the project manager, lest the contingency funds be consumed before future needs materialize. The plot presented in Fig. A2-2 is a plot of only columns G and H of Table A2-5. It plots both dollar amount and percentage consumed against percentage completion.

This is the end of the contingency evaluation and drawdown identification process at budgeting stages. The actual execution of construction may yield differences from what was budgeted, but detailed cost control is the subject of another book.

Table A2-5. Contingency Fund Allocation

Work package	Underrun ($ million) (A)	Overrun ($ million) (B)	Probability of underrun (C)	Probability of overrun (D)	Expected overrun-expected underrun ($ million) (E)	% Allocation (F)	Contingency fund allocation = F × contingency amount ($ million) (G)	% Complete (H)
Utilities	$1	$1	0.2	0.4	$(1 \times 0.4) - (1 \times 0.2) =$ $0.2	$0.2/3.6 =$ 5.56%	$0.0556 \times \$7 = \$0.39$	20%
Concrete structure	$1	$3	0.2	0.4	$(3 \times 0.4) - (1 \times 0.2) =$ $1.0	$1.0/3.6 =$ 27.78%	$0.2778 \times \$7 = \$1.94$	45%
Steel	$1	$2	0.2	0.5	$(2 \times 0.5) - (1 \times 0.2) =$ $0.8	$0.8/3.6 =$ 22.22%	$0.2222 \times \$7 = \$1.56$	95%
Interior works	$1	$3	0.2	0.6	$(3 \times 0.6) - (1 \times 0.2) =$ $1.6	$1.6/3.6 =$ 44.44%	$0.4444 \times \$7 = \$3.11$	100%
Total					$3.60	100%	$7.00	

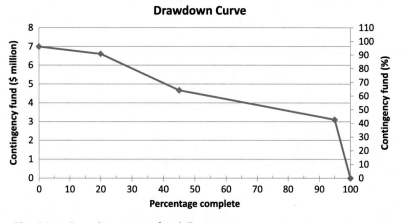

Fig. A2-2. Drawdown curve for dollar amount

Index